LECTURE NOTES ON
LOCAL RINGS

LECTURE NOTES ON
LOCAL RINGS

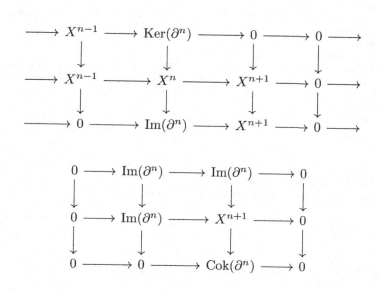

Birger Iversen

edited by
Holger Andreas Nielsen
Aarhus University, Denmark

World Scientific

NEW JERSEY · LONDON · SINGAPORE · BEIJING · SHANGHAI · HONG KONG · TAIPEI · CHENNAI

Published by

World Scientific Publishing Co. Pte. Ltd.

5 Toh Tuck Link, Singapore 596224

USA office: 27 Warren Street, Suite 401-402, Hackensack, NJ 07601

UK office: 57 Shelton Street, Covent Garden, London WC2H 9HE

Library of Congress Cataloging-in-Publication Data
Iversen, Birger, author.
 [Lecture notes. Selections]
 Lecture notes on local rings / by Birger Iversen ; edited by Holger Andreas Nielsen (Aarhus University, Denmark).
   pages cm
 Includes bibliographical references and index.
   ISBN 978-9814603652 (hardcover : alk. paper) -- ISBN 9814603651 (hardcover : alk. paper) --
1. Local rings. 2. Injective modules (Algebra) 3. Intersection homology theory. I. Nielsen, Holger Andreas, editor of compilation. II. Title.
   QA251.38.I94 2014
   512'.4--dc23
                                                          2014010375

British Library Cataloguing-in-Publication Data
A catalogue record for this book is available from the British Library.

In-house Editor: Elizabeth Lie

Printed in Singapore

Preface

Professor Iversen wrote the notes on local rings over the years 1974–80. The influence of Bourbaki, Grothendieck and Serre is obvious. Since then the notes have been a basis for several courses on local rings at the University of Aarhus. Even though a lot has happened in commutative algebra in the last 20 years, these notes contain material and viewpoints that are not so easy to find in a collected form in literature.

The content in Chapter 1–3 is a fairly standard one–semester course on local rings with the goal to reach the fact that a regular local ring is a unique factorization domain. The homological machinery is also supported by Cohen–Macaulay rings and depth. In Chapters 4–6 the methods of injective modules, Matlis duality and local cohomology are discussed. Chapters 7–9 are not so standard and introduce the reader to the generalizations of modules to complexes of modules. Some of Professor Iversen's research results are given in Chapter 9. Chapter 10 is about Serre's intersection conjecture. The graded case is fully exposed. The last chapter introduces the reader to Fitting ideals and McRae invariants.

In June 1995 the last time I met Professor Iversen, we discussed finishing the notes on local rings. Several points were not satisfying. An introductory chapter on module theory, tensor product and simple homological algebra was missing. Example material including the polynomial rings should be incorporated. Professor Iversen had some ideas on including Cohen's structure theorem. Some elaboration on divisor theory was in mind. A set of exercises for each chapter should be collected from the past years' courses.

Now after thinking about the project for 7 years, I must confess that the goal is not in sight. So I have decided to bring the notes forward remixed close to the original form, but with some redactional changes. The bibliography is quite an informal list of relevant material Professor Iversen was aware of.

At last, note that all rings are commutative with 1 and all modules are unitary.

Holger Andreas Nielsen, University of Aarhus, Summer 2002

Added January 2014

This project was brought alive in 2013 by the initiative of Professor Niels Lauritzen. The beautiful typesetting is a result of the efforts of Lars Madsen.

Aarhus University, 2014

Contents

Chapter 1

Dimension of a Local Ring

1.1 Nakayama's lemma

Recall that an ideal $\mathfrak{m}$ in a ring A is called a *maximal ideal* if $\mathfrak{m} \neq A$ and is maximal among the ideals in A different from A, or what amounts to the same, if $A/\mathfrak{m}$ is a field.

Theorem 1.1 (Krull). *Any ring A which is different from zero contains a maximal ideal.*

Proof. The nonempty set of ideals in A different from A is easily seen to be inductively ordered (by inclusion). Conclusion by Zorn's lemma. $\qquad\square$

Definition 1.2. A *local ring* is a ring A which contains precisely one maximal ideal. The residue field of the local ring A with respect to its maximal ideal is called *the residue field* of A.

Let A be a local ring with maximal ideal $\mathfrak{m}$. It follows immediately from 1.1 that the elements of A outside $\mathfrak{m}$ are invertible. Conversely, suppose given a ring A and a maximal ideal $\mathfrak{m}$ such that all elements of A outside $\mathfrak{m}$ are invertible, then A is a local ring.

Lemma 1.3 (Nakayama's lemma). *Let A be a local ring with maximal ideal $\mathfrak{m}$ and M a finitely generated A-module. Then $M/\mathfrak{m}M = 0$ implies $M = 0$.*

Proof. We shall give two proofs of this lemma.

Suppose $M \neq 0$ is a finitely generated A-module. Note that the set of proper submodules of M is inductively ordered and whence by Zorn's lemma contains a maximal element, say, N. The module M/N being a simple module is annihilated by $\mathfrak{m}$, whence $\mathfrak{m}M \subseteq N \neq M$.

1

For the second proof let $x_1, \ldots, x_d \in M$ generate M and suppose $M = \mathfrak{m}M$. We can find $a_{ij} \in \mathfrak{m}$ such that

$$x_i = \sum_j a_{ij} x_j$$

or

$$0 = \sum_j (\delta_{ij} - a_{ij}) x_j$$

where δ_{ij} is a Kronecker symbol. Note that the $d \times d$-matrix $(\delta_{ij} - a_{ij})$ has determinant congruent to 1 mod $\mathfrak{m}$. 1.1 makes this determinant invertible and consequently $(\delta_{ij} - a_{ij})$ is an invertible matrix. $\qquad\square$

1.2   Prime ideals

Let A be a ring, recall that a *prime ideal* in A is an ideal $\mathfrak{p}$, distinct from A such that

$$ab \in \mathfrak{p} \quad \Rightarrow \quad a \in \mathfrak{p} \text{ or } b \in \mathfrak{p}, \qquad a, b \in A.$$

It is the same to say that $A/\mathfrak{p}$ is an integral domain. $A_{\mathfrak{p}}$ denotes the *localized* ring where all elements outside $\mathfrak{p}$ are inverted. The maximal ideal is $\mathfrak{p}A_{\mathfrak{p}}$ and the *residue field* $k_{\mathfrak{p}}$ is the fraction field of $A/\mathfrak{p}$. If M is an A-module then $M_{\mathfrak{p}}$ denotes the localized $A_{\mathfrak{p}}$-module.

Remark 1.4. Let $\mathfrak{p}$ be a prime ideal in A, $\mathfrak{a}$ and $\mathfrak{b}$ ideals in A such that $\mathfrak{a}\mathfrak{b} \subseteq \mathfrak{p}$. Then $\mathfrak{a} \subseteq \mathfrak{p}$ or $\mathfrak{b} \subseteq \mathfrak{p}$, as one readily verifies.

Proposition 1.5. *Let all but at most two $\mathfrak{p}_1, \ldots, \mathfrak{p}_d$ be prime ideals in the ring A and $\mathfrak{a}$ an ideal in A. If*

$$\mathfrak{a} \subseteq \bigcup_i \mathfrak{p}_i$$

then $\mathfrak{a} \subseteq \mathfrak{p}_i$ for some i.

Proof. Induction on $d \geq 2$. Suppose there exists j with $1 \leq j \leq d$ such that

$$\mathfrak{a} \cap \mathfrak{p}_j \subseteq \bigcup_{i \neq j} \mathfrak{p}_i.$$

Since $\mathfrak{a} = \bigcup_i (\mathfrak{a} \cap \mathfrak{p}_i)$, this implies $\mathfrak{a} \subseteq \bigcup_{i \neq j} \mathfrak{p}_i$ and the statement follows from the induction hypothesis. Thus we may choose for each $1 \leq j \leq d$

an element $y_j \in \mathfrak{a} \cap \mathfrak{p}_j$ such that $y_j \notin \mathfrak{p}_i$ for $i \neq j$. Consider the following element of $\mathfrak{a}$,

$$y_d + \prod_{i<d} y_i.$$

As is easily seen, this element does not belong to any of the $\mathfrak{p}_j$'s, $1 \leq j \leq d$; this is a contradiction to the assumption in the proposition. Let us remark that the last assertion uses that $\mathfrak{p}_d$ is a prime only if $d > 2$. □

Proposition 1.6. *Let* $\mathfrak{p}_1, \ldots, \mathfrak{p}_d$ *be prime ideals in the ring* A *and* $\mathfrak{a}$ *an ideal in* A. *If for some* $y \in A$

$$y + \mathfrak{a} \subseteq \bigcup_i \mathfrak{p}_i$$

then $\mathfrak{a} \subseteq \mathfrak{p}_i$ *for some* i.

Proof. If $y \in \bigcap_i \mathfrak{p}_i$ then conclusion by 1.5. On the contrary after renumbering there exists j with $1 \leq j < d$ such that

$$y \in \mathfrak{p}_1 \cap \cdots \cap \mathfrak{p}_j \quad \text{and} \quad y \notin \mathfrak{p}_{j+1} \cup \cdots \cup \mathfrak{p}_d.$$

Assume no inclusions between the prime ideals and choose by 1.4

$$x \in \mathfrak{a} \cap \mathfrak{p}_{j+1} \cap \cdots \cap \mathfrak{p}_d \quad \text{and} \quad x \notin \mathfrak{p}_1 \cup \cdots \cup \mathfrak{p}_j.$$

Then $y + x \notin \bigcup_i \mathfrak{p}_i$ contradicts the hypothesis. □

Proposition 1.7. *Let* A *be a ring. Then the intersection of all prime ideals in* A *equals the set of nilpotent elements in* A.

Proof. A nilpotent element $f \in A$ is clearly contained in any prime ideal of A. Suppose $f \in A$ is not nilpotent, by Zorn's lemma we get an ideal maximal among the ideals not containing any power of f. This is a prime ideal not containing f. □

Proposition 1.8. *Let* A *be a ring. Any prime ideal of* A *contains a minimal prime ideal (i.e., a prime ideal which is minimal in the set of prime ideals in* A *ordered by inclusion).*

Proof. The set of prime ideals in A is easily seen to be (downward) inductively ordered. Conclusion by Zorn's lemma. □

1.3   Noetherian modules

In this section we are going to introduce the basic finiteness condition in the topic.

Proposition 1.9. *Let M be a module over the ring A. The following three conditions are equivalent*

(1) *Any submodule of M is finitely generated.*

(2) *Any increasing sequence $(M_i)_{i \in \mathbb{N}}$ of submodules is stationary (i.e., $M_i = M_{i+1}$ for $i \gg 0$).*

(3) *Any nonempty subset of submodules of M contains a maximal element.*

Proof. $(1) \Rightarrow (2)$: The union N of the M_i's is easily seen to be a submodule of M. Let $x_1, \dots, x_d$ generate that submodule. We have $x_1, \dots, x_d \in M_i$ for $i \gg 0$. This makes the sequence stationary.

$(2) \Rightarrow (3)$: Suppose $\mathcal{M}$ is a nonempty subset of submodules of M which do not contain a maximal element. A module M_i in $\mathcal{M}$ is strictly contained in some module M_{i+1} from $\mathcal{M}$. Continuing this process gives a strictly increasing sequence of submodules of M in contradiction to the hypothesis.

$(3) \Rightarrow (1)$: Let N be a submodule of M and let $\mathcal{N}$ denote the set of submodules of N which are finitely generated. Let N' be a maximal element in that set. For any $x \in N$, $Ax + N' \in \mathcal{N}$, whence $Ax + N' = N'$, i.e., $x \in N'$. Thus $N = N'$ and so $N \in \mathcal{N}$ is finitely generated.   □

Definition 1.10. A module M over a ring A which satisfies the conditions of 1.9 is called a *noetherian module*. The ring A itself is called a *noetherian ring* if A is a noetherian module over itself.

Lemma 1.11. *Let $0 \to N \to M \to P \to 0$ be an exact sequence of modules over the ring A. Then M is noetherian if and only if N and P are noetherian.*

Proof. If M is noetherian then submodules of N and P are finitely generated, so N and P are noetherian by 1.9. Conversely, if N and P are noetherian and $M' \subseteq M$ is a submodule, then the induced submodules $N' \subseteq N$ and $P' \subseteq P$ are finitely generated. From the induced exact sequence $0 \to N' \to M' \to P' \to 0$ it follows that M' is finitely generated, so M is noetherian by 1.9.   □

Proposition 1.12. *Let A be a noetherian ring. Any finitely generated A-module is noetherian.*

Proof. Let M be a finitely generated A-module. We can find a surjective morphism $A^n \to M \to 0$ for some $n \in \mathbb{N}$. Conclusion by 1.11. $\qquad \square$

Theorem 1.13. *Let A be a noetherian ring and $M \neq 0$ a finitely generated A-module. Then there exists a finite filtration of M by submodules*

$$0 = M_0 \subset M_1 \subset \cdots \subset M_{r-1} \subset M_r = M$$

such that M_i/M_{i-1}, $i = 1, \ldots, r$ is isomorphic to an A-module of the form $A/\mathfrak{p}_i$ where $\mathfrak{p}_i$ is a prime ideal in A.

Proof. Let us first prove that if $P \neq 0$ is an A-module, then there exists an element $x \in P$ whose annihilator (the set of elements $a \in A$, such that $ax = 0$), $\mathrm{Ann}(x)$ is a prime ideal. Consider the set $\mathcal{P}$ of ideals in A which are of the form $\mathrm{Ann}(x)$, for some $x \in P$, $x \neq 0$. Choose $x \in P$, $x \neq 0$, such that $\mathrm{Ann}(x)$ is maximal in $\mathcal{P}$. We are going to prove that $\mathrm{Ann}(x)$ is a prime ideal. Let $a, b \in A$ be such that $ab \in \mathrm{Ann}(x)$ and suppose $b \notin \mathrm{Ann}(x)$. We have

$$\mathrm{Ann}(x) \subseteq \mathrm{Ann}(bx) \neq A.$$

Consequently $\mathrm{Ann}(x) = \mathrm{Ann}(bx)$ in particular $a \in \mathrm{Ann}(x)$. Let now $\mathcal{M}$ denote the set of submodules of M for which the theorem is true (i.e., each of which admits a filtration...). Then $\mathcal{M} \neq \emptyset$ as we have just proved. Let $N \in \mathcal{M}$ be maximal in this set. We are going to prove that $N = M$. Suppose $N \neq M$. By the first result applied to M/N we can find a submodule N' with $N \subset N' \subseteq M$, such that N'/N is isomorphic to an A-module of the form $A/\mathfrak{p}'$ where $\mathfrak{p}'$ is a prime ideal; this is easily seen to contradict the maximality of N in $\mathcal{M}$. $\qquad \square$

Proposition 1.14. *Let A be a noetherian ring. The number of minimal prime ideals is finite.*

Proof. Choose a filtration

$$0 = \mathfrak{a}_0 \subset \mathfrak{a}_1 \subset \cdots \subset \mathfrak{a}_{r-1} \subset \mathfrak{a}_r = A$$

of A by ideals such that each $\mathfrak{a}_i/\mathfrak{a}_{i-1}$ is isomorphic to an A-module of the form $A/\mathfrak{p}_i$, where $\mathfrak{p}_i$ is a prime ideal. Let $\mathfrak{p}$ be a minimal prime ideal in A. We are going to prove that $\mathfrak{p} = \mathfrak{p}_i$ for some i, $1 \leq i \leq r$. Note that $\mathfrak{p} \neq \mathfrak{p}_i$ is equivalent to the localization $(\mathfrak{a}_i/\mathfrak{a}_{i-1})_\mathfrak{p} = 0$ since $\mathfrak{p}$ is minimal. Thus $\mathfrak{p} \neq \mathfrak{p}_i$ for all i implies $A_\mathfrak{p} = 0$, which is a contradiction. $\qquad \square$

1.4 Modules of finite length

Let A be a ring. An A-module M is said to be *of finite length* if it admits a filtration by submodules

$$0 = M_0 \subset M_1 \subset \cdots \subset M_{r-1} \subset M_r = M$$

such that each M_i/M_{i-1}, $i = 1, \ldots, r$ is a *simple module* (a nonzero module which admits no nontrivial submodules).

Lemma 1.15. *Any two finite filtrations with simple quotients have the same number of submodules.*

Proof. Let $l(M)$ be the least length of a filtration of M. If M is simple then $l(M) = 1$. Look at a filtration

$$0 = M_0 \subset M_1 \subset \cdots \subset M_r = M$$

with simple quotients M_i/M_{i-1}. For submodule $N \subseteq M$ we have a filtration

$$0 = N \cap M_0 \subseteq N \cap M_1 \subseteq \cdots \subseteq N \cap M_r = N$$

with quotients $N \cap M_i / N \cap M_{i-1} \subseteq M_i/M_{i-1}$ being either simple or 0. It follows that

$$N \subset M \quad \Rightarrow \quad l(N) < l(M).$$

Applying this to the filtration above gives

$$l(M) \leq r \leq l(M)$$

as wanted. $\square$

The number of nontrivial submodules in a filtration as above will be denoted $\ell_A(M)$ and is called *the length* of M. A submodule N of the module M is said to be of finite *colength* if M/N has finite length.

Proposition 1.16. *Given an exact sequence*

$$0 \longrightarrow N \longrightarrow M \longrightarrow P \longrightarrow 0.$$

Then M has a finite length if and only if both N and P have a finite length. In that case

$$\ell_A(M) = \ell_A(N) + \ell_A(P).$$

Proof. A finite filtration with simple quotients in M induces filtrations in both N and P, and vice versa.

The filtrations in N and P give a filtration in M with the same quotients, so the length formula follows from 1.15. □

Definition 1.17. A ring A is called an *artinian ring* if A is a module of finite length over itself.

Note that an artinian ring is noetherian and that any finitely generated A-module has finite length.

Proposition 1.18. *Let A be a noetherian ring. Then A is an artinian ring if and only if all prime ideals are maximal ideals.*

Proof. Suppose A is an artinian ring and let $\mathfrak{p}$ be a prime ideal in A. Then $A/\mathfrak{p}$ is a field since by 1.16 one deduces that an injective endomorphism of a module of finite length is an isomorphism. Conversely, suppose all prime ideals of A are maximal, then all finitely generated A-modules have finite length by 1.13. □

Let A be an artinian ring. Consider a filtration of A with simple quotients to see that A has only finitely many maximal ideals (this may also be deduced from 1.14 combined with 1.18). More precisely

Proposition 1.19. *Any artinian ring is the product of finitely many artinian local rings.*

Proof. Let $\mathfrak{m}_1, \ldots, \mathfrak{m}_d$ denote the maximal ideals and let n_i denote the number of factors in a filtration of A with simple quotients, which are isomorphic to $A/\mathfrak{m}_i$. Each $n_i \geq 1$. Again by considering the filtration of A it is easily seen that the A-module A is annihilated by $\prod \mathfrak{m}_i^{n_i}$, i.e.,

$$\prod_i \mathfrak{m}_i^{n_i} = 0.$$

Conclusion by Chinese remainder theorem, 1.20 below. □

Theorem 1.20 (Chinese remainder theorem). *Let A denote a ring and $\mathfrak{a}_1, \ldots, \mathfrak{a}_d$ a family of ideals such that $\mathfrak{a}_i + \mathfrak{a}_j = A$ for all $i \neq j$. Then the canonical map*

$$A / \prod_i \mathfrak{a}_i \to \prod_i A/\mathfrak{a}_i$$

is an isomorphism.

Proof. Induction on d. In case $d = 2$ we have the exact sequence

$$0 \longrightarrow A/\mathfrak{a}_1 \cap \mathfrak{a}_2 \longrightarrow A/\mathfrak{a}_1 \times A/\mathfrak{a}_2 \longrightarrow A/\mathfrak{a}_1 + \mathfrak{a}_2 \longrightarrow 0.$$

Thus it suffices to prove that $\mathfrak{a}_1\mathfrak{a}_2 = \mathfrak{a}_1 \cap \mathfrak{a}_2$ under the assumption $\mathfrak{a}_1 + \mathfrak{a}_2 = A$. Let $e_1 \in \mathfrak{a}_1$ and $e_2 \in \mathfrak{a}_2$ be such that $e_1 + e_2 = 1$. For an element $a \in \mathfrak{a}_1 \cap \mathfrak{a}_2$ we have $a = ae_1 + ae_2 \in \mathfrak{a}_1\mathfrak{a}_2$. In the general case put $\mathfrak{a}_2' = \prod_{i>1} \mathfrak{a}_i$, it follows from 1.1 that $\mathfrak{a}_1 + \mathfrak{a}_2' = A$. Whence by our previous result

$$A/\prod \mathfrak{a}_i \simeq A/\mathfrak{a}_1 \times A/\mathfrak{a}_2'$$

and by the induction hypothesis

$$A/\mathfrak{a}_2' \simeq \prod_{i>1} A/\mathfrak{a}_i$$

concluding the proof. $\qquad\square$

1.5 Hilbert's basis theorem

The following theorem is fundamental and gives rise to the basic examples.

Theorem 1.21 (Hilbert's basis theorem). *Let A be a noetherian ring. Then $A[X]$, the ring of polynomials with coefficients in A, is noetherian.*

Proof. Let $\mathfrak{I} \subseteq A[X]$ be an ideal and let $\mathfrak{a}$ denote the ideal consisting of the leading coefficients of the elements of $\mathfrak{I}$. Let $a_1, \ldots, a_s$ generate the ideal $\mathfrak{a}$ and for each $1 \leq i \leq s$ choose $f_i \in \mathfrak{I}$ such that

$$f_i = a_i X^{d_i} + \text{terms of lower degree.}$$

Put $d = \sup_i d_i$ and let $A[X]_{\leq d}$ denote the A-module of polynomials of degree $\leq d$. We leave it to the reader to prove that

$$\mathfrak{I} = \mathfrak{I} \cap A[X]_{\leq d} + (f_1, \ldots, f_s),$$

where $(f_1, \ldots, f_s)$ denotes the ideal in $A[X]$ generated by $f_1, \ldots, f_s$. It follows that the A-module $\mathfrak{I} \cap A[X]_{\leq d}$ is finitely generated, being a sub A-module of the finitely generated A-module $A[X]_{\leq d}$. If $f_{s+1}, \ldots, f_t$ generate that A-module, then $f_1, \ldots, f_t$ generate the ideal $\mathfrak{I}$. $\qquad\square$

Corollary 1.22. *The rings $\mathbb{Z}[X_1, \ldots, X_n]$, $\mathbb{C}[X_1, \ldots, X_n], \ldots$ are noetherian rings.*

1.6 Graded rings

By a *graded ring* A we understand a family $(A_i)_{i \in \mathbb{N}}$ of abelian groups and a structure of ring on $\bigoplus_{i \in \mathbb{N}} A_i$ such that

$$A_i A_j \subseteq A_{i+j}, \qquad i, j \in \mathbb{N}.$$

It follows that $1 \in A_0$, that A_0 is a subring and that A_i is an A_0-module, $i \in \mathbb{N}$.

By a *graded module* M over a graded ring A we understand a family $(M_i)_{i \in \mathbb{Z}}$ of abelian groups and a structure of $\bigoplus_{i \in \mathbb{N}} A_i$-module on $\bigoplus_{i \in \mathbb{Z}} M_i$, such that $A_i M_j \subseteq M_{i+j}$. Let M and N be graded modules over a graded ring A. By a *morphism* $f : M \to N$ we understand a morphism of the underlying modules such that $f(M_i) \subseteq N_i$. Let M denote a graded module over the graded ring A and $s \in \mathbb{Z}$. Let $M[s]$ denote the graded module whose underlying module is the same as that of M, but the grading is given by $(M_{i+s})_{i \in \mathbb{Z}}$, i.e., $M[s]_i = M_{i+s}$.

Proposition 1.23. *Let A denote a graded ring such that the ideal $A_+ = \bigoplus_{i \geq 1} A_i$ is generated by A_1. If A_0 is noetherian and A_1 is a finitely generated A_0-module, then A is noetherian.*

Proof. Let $x_1, \ldots, x_d$ generate the A_0-module A_1 and consider the morphism of A_0-rings

$$A_0[X_1, \ldots, X_d] \to A, \qquad X_i \to x_i.$$

It is easy to verify that the specific assumption made on A makes this morphism surjective. Conclusion by Hilbert's basis theorem, 1.21. □

Proposition 1.24. *Let M denote finitely generated graded module over the noetherian graded ring A. Then M_i is a finitely generated A_0-module. Moreover $M_i = 0$ for $i \ll 0$.*

Proof. Consider the graded submodule

$$M_{|i} = \bigoplus_{t \geq i} M_t.$$

This is a finitely generated submodule, and whence $M_i \simeq M_{|i}/M_{|i+1}$ is a finitely generated A_0-module. The last statement is clear. □

In the rest of this section we shall consider graded modules over a polynomial ring which is considered graded in the standard way by degree.

Theorem 1.25. *Let k denote an artinian ring and let $A = k[X_0, \ldots, X_n]$ be the polynomial ring in $n + 1$ variables. For any finitely generated graded A-module M, there exists $P \in \mathbb{Q}[T]$ such that*

$$P(i) = \ell_k(M_i), \qquad i \gg 0.$$

Moreover this polynomial has degree $\leq n$.

Proof. Induction on n. If $n = -1$ we have $M_i = 0$ for $i \gg 0$. Suppose $n \geq 0$. Consider the exact sequence

$$0 \longrightarrow K \longrightarrow M \overset{\cdot X_n}{\longrightarrow} M[1] \longrightarrow C \longrightarrow 0.$$

Note that K and C are annihilated by X_n and consequently are finitely generated modules over the subring $k[X_0, \ldots, X_{n-1}]$ of A. We get

$$\ell_k(M_{i+1}) - \ell_k(M_i) = \ell_k(C_i) - \ell_k(K_i)$$

and consequently the induction hypothesis implies that there exists a polynomial R with rational coefficients of degree $\leq n - 1$ such that

$$\ell_k(M_{i+1}) - \ell_k(M_i) = R(i), \qquad i \gg 0.$$

By 1.26 below we can find $P \in \mathbb{Q}[T]$ such that

$$\ell_k(M_{i+1}) - \ell_k(M_i) = P(i+1) - P(i), \qquad i \gg 0.$$

From this the conclusion follows easily.   □

Lemma 1.26. *For $P \in \mathbb{Q}[T]$, put $\Delta P = P(T+1) - P(T)$. The map*

$$\Delta : \mathbb{Q}[T] \to \mathbb{Q}[T]$$

is surjective.

Proof. For $r \in \mathbb{N}$ put

$$\binom{T}{r} = \frac{T(T-1) \cdots (T - r + 1)}{r!}, \qquad \binom{T}{0} = 1.$$

Note that $\Delta\binom{T}{r} = \binom{T}{r-1}$, $r \geq 1$.   □

Remark 1.27. Let $P \in \mathbb{Q}[T]$. Suppose $P(d) \in \mathbb{Z}$ for $d \in \mathbb{N}$, $d \gg 0$, then

$$P(T) = \sum_{i=0}^{m} e_i \binom{T}{i}, \qquad e_i \in \mathbb{Z}.$$

Proof. Induction on degree of P. Use that $\Delta\binom{T}{r} = \binom{T}{r-1}$, $r \geq 1$. $\qquad\square$

Definition 1.28. With the notation of 1.25, we let the *Hilbert polynomial* $\mathfrak{H}(M, T)$ be the polynomial with rational coefficients such that

$$\mathfrak{H}(M, i) = \ell_k(M_i), \qquad i \gg 0.$$

Example 1.29. Let k denote a field.

$$\mathfrak{H}(k[X_0, \ldots, X_n], T) = \binom{T + n}{n}.$$

Example 1.30. Let k denote a field and $H \in k[X_0, \ldots, X_n]$ a homogeneous polynomial of degree d.

$$\mathfrak{H}(k[X_0, \ldots, X_n]/(H), T) = \binom{T + n}{n} - \binom{T + n - d}{n}.$$

From this one deduces that for $0 \neq \mathfrak{I} \subseteq k[X_0, \ldots, X_n]$ a graded ideal,

$$\deg(\mathfrak{H}(k[X_0, \ldots, X_n]/\mathfrak{I}, T)) \leq n - 1.$$

1.7 Filtered rings

By a *filtered ring* A we understand a ring A^0 and a descending filtration $(A^i)_{i \in \mathbb{N}}$, $A^{i+1} \subseteq A^i$ of A^0 by ideals such that $A^i A^j \subseteq A^{i+j}$, $i, j \in \mathbb{N}$. A *filtered module* M over the filtered ring A is an A^0-module M^0 and a descending filtration $(M^i)_{i \in \mathbb{N}}$, $M^{i+1} \subseteq M^i$ of M^0 by submodules such that $A^i M^j \subseteq M^{i+j}$, $i, j \in \mathbb{N}$. Given filtered modules M, N over the filtered ring A. By a *morphism* $f : M \to N$ we understand an A^0-linear map $f : M^0 \to N^0$ such that $f(M^i) \subseteq N^i$.

To a filtered ring we associate a graded ring

$$\mathrm{gr}(A) = \bigoplus_{i \in \mathbb{N}} A^i/A^{i+1},$$

where multiplication is induced by multiplication $A^i \times A^j \to A^{i+j}$. Similarly to a filtered module M we associate a graded $\mathrm{gr}(A)$-module

$$\mathrm{gr}(M) = \bigoplus_{i \in \mathbb{N}} M^i / M^{i+1},$$

where the module structure is induced from the structure map $A^i \times M^j \to M^{i+j}$. A morphism $f : M \to N$ of filtered modules induces in an obvious way a morphism of graded $\mathrm{gr}(A)$-modules

$$\mathrm{gr}(f) : \mathrm{gr}(M) \to \mathrm{gr}(N).$$

Example 1.31. Let A be a ring and $\mathfrak{m}$ an ideal in A. The filtration $(\mathfrak{m}^i)_{i \in \mathbb{N}}$ of A is called the $\mathfrak{m}$-*adic filtration*. The graded ring associated to this filtered ring is denoted $\mathrm{gr}_{\mathfrak{m}}(A)$. Similarly given an A-module M the filtration $(\mathfrak{m}^i M)_{i \in \mathbb{N}}$ is called the $\mathfrak{m}$-*adic filtration* of M. The associated graded module is denoted by $\mathrm{gr}_{\mathfrak{m}}(M)$.

Definition 1.32. Let A be a ring, $\mathfrak{m}$ an ideal in A and M an A-module. A filtration $(M^i)_{i \in \mathbb{N}}$ of M is called an $\mathfrak{m}$-*filtration* if

$$\mathfrak{m}M^i \subseteq M^{i+1}, \qquad i \in \mathbb{N}.$$

An $\mathfrak{m}$-filtration $(M^i)_{i \in \mathbb{N}}$ of M is called *stable* if

$$\mathfrak{m}M^i = M^{i+1}, \qquad i \gg 0.$$

Lemma 1.33 (Artin–Rees' lemma). *Let A be a noetherian ring, $\mathfrak{m}$ an ideal in A, M a finitely generated A-module, and N a submodule of M. Then any stable $\mathfrak{m}$-filtration $(M^i)_{i \in \mathbb{N}}$ of M induces $(N^i = N \cap M^i)_{i \in \mathbb{N}}$ a stable $\mathfrak{m}$-filtration of N.*

Proof. We shall first give a criterion for stability for an arbitrary $\mathfrak{m}$-filtration $(M^i)_{i \in \mathbb{N}}$ of M. Consider the graded ring

$$\bar{A} = \bigoplus_{i \in \mathbb{N}} \mathfrak{m}^i$$

and the graded $\bar{A}$-module

$$\bar{M} = \bigoplus_{i \in \mathbb{N}} M^i,$$

where the ring and module structure respectively are the obvious ones. We shall prove that the filtration of M is stable if and only if $\bar{M}$ is a finitely generated $\bar{A}$-module.

For this we shall introduce an ascending sequence $(\bar{M}_t)_{t \in \mathbb{N}}$ of submodules of $\bar{M}$: define

$$\bar{M}_t = M^0 \oplus \cdots \oplus M^t \oplus \mathfrak{m}M^t \oplus \mathfrak{m}^2 M^t \oplus \cdots.$$

Note that the filtration of M is stable if and only if $\bar{M}_t = \bar{M}$ for some t. On the other hand, note that $\bar{M}_t$ is a finitely generated $\bar{A}$-module and that $\bar{M} = \bigcup_{t \in \mathbb{N}} \bar{M}_t$. Suppose M is finitely generated, then $\bar{M}$ is noetherian since $\bar{A}$ is noetherian as it follows from 1.23, and whence $(\bar{M}_t)_{t \in \mathbb{N}}$ is stationary, i.e., $\bar{M}_t = \bar{M}$ for some t.

To conclude the proof, note that $\bar{N}$ is a submodule of $\bar{M}$, stability of the $\mathfrak{m}$-filtration $(M^i)_{i \in \mathbb{N}}$ implies that $\bar{M}$ and whence $\bar{N}$ is finitely generated, i.e., $(N^i = N \cap M^i)_{i \in \mathbb{N}}$ is a stable $\mathfrak{m}$-filtration of N. $\qquad\square$

Corollary 1.34. *Let A be a noetherian ring, $\mathfrak{m}$ an ideal in A, M a finitely generated A-module, and N a submodule of M. Then there exists an $i_0 \in \mathbb{N}$ such that*

$$N \cap \mathfrak{m}^i M = \mathfrak{m}^{i-i_0}(N \cap \mathfrak{m}^{i_0} M), \qquad i \geq i_0.$$

Proof. This is nothing but Artin–Rees' lemma, 1.33, applied to the $\mathfrak{m}$-adic filtration of M. $\qquad\square$

1.8 Local rings

Throughout this section A denotes a noetherian local ring, $\mathfrak{m}$ its maximal ideal and $k = A/\mathfrak{m}$ the residue field of A.

Theorem 1.35 (Krull's intersection theorem). *Let M denote a finitely generated A-module. Then*

$$\bigcap_{i \in \mathbb{N}} \mathfrak{m}^i M = 0.$$

Proof. Put $N = \bigcap_{i \in \mathbb{N}} \mathfrak{m}^i M$. By Artin–Rees' lemma, 1.33,

$$N \cap \mathfrak{m}^i M = \mathfrak{m}^1(N \cap \mathfrak{m}^{i-1}M), \qquad i \gg 0.$$

i.e., $N = \mathfrak{m}N$, whence $N = 0$ by Nakayama's lemma, 1.3. $\qquad\square$

Proposition 1.36. *Let M denote a finitely generated A-module. Then there exists $P \in \mathbb{Q}[T]$ such that*

$$\ell_A(M/\mathfrak{m}^i M) = P(i), \qquad i \gg 0.$$

Proof. Consider the graded $\mathrm{gr}_{\mathfrak{m}}(A)$-module $\mathrm{gr}_{\mathfrak{m}}(M)$. Choose a basis for the k-vector space $\mathfrak{m}/\mathfrak{m}^2$ to obtain a surjection of graded rings

$$k[X_1, \ldots, X_n] \to \mathrm{gr}_{\mathfrak{m}}(A).$$

Thus it follows from 1.25 that $\ell_k([\mathrm{gr}_{\mathfrak{m}}(M)]_i)$ is a rational polynomial in i for $i \gg 0$. Note that

$$\ell_A(M/\mathfrak{m}^{i+1}M) - \ell_A(M/\mathfrak{m}^i M) = \ell_k([\mathrm{gr}_{\mathfrak{m}}(M)]_i)$$

and conclude by 1.26. $\qquad\square$

Definition 1.37. Let M denote a finitely generated A-module. Let the *Samuel polynomial* $\chi_{\mathfrak{m}}(M, T) \in \mathbb{Q}[T]$ be the polynomial such that

$$\chi_{\mathfrak{m}}(M, i) = \ell_A(M/\mathfrak{m}^i M), \qquad i \gg 0.$$

Note, $\Delta\chi_{\mathfrak{m}}(M, T) = \mathfrak{H}(\mathrm{gr}_{\mathfrak{m}}(M), T)$.

Proposition 1.38. *Consider a strictly ascending chain of prime ideals in A:*

$$\mathfrak{p}_0 \subset \mathfrak{p}_1 \subset \cdots \subset \mathfrak{p}_r.$$

*Hereafter called **a chain of prime ideals of length** r. Then*

$$r \leq \deg(\chi_{\mathfrak{m}}(A, T)).$$

Proof. As is easily seen we have

$$0 \leq \deg(\chi_{\mathfrak{m}}(A/\mathfrak{p}_r, T)) \leq \cdots \leq \deg(\chi_{\mathfrak{m}}(A/\mathfrak{p}_0, T)) \leq \deg(\chi_{\mathfrak{m}}(A, T)).$$

Thus it suffices to prove that

$$\deg(\chi_{\mathfrak{m}}(A/\mathfrak{p}_{i+1}, T)) < \deg(\chi_{\mathfrak{m}}(A/\mathfrak{p}_i, T)), \qquad i = 0, 1, \ldots, r-1.$$

Choose $a \in \mathfrak{p}_{i+1} - \mathfrak{p}_i$ and consider the short exact sequence

$$0 \longrightarrow A/\mathfrak{p}_i \xrightarrow{\ a\ } A/\mathfrak{p}_i \longrightarrow A/(a, \mathfrak{p}_i) \longrightarrow 0.$$

By 1.39 below we get

$$\deg(\chi_{\mathfrak{m}}(A/(a, \mathfrak{p}_i), T)) < \deg(\chi_{\mathfrak{m}}(A/\mathfrak{p}_i, T)),$$

but clearly

$$\deg(\chi_{\mathfrak{m}}(A/\mathfrak{p}_{i+1}, T)) \leq \deg(\chi_{\mathfrak{m}}(A/(a, \mathfrak{p}_i), T)). \qquad\square$$

Lemma 1.39. *Given a short exact sequence of finitely generated A-modules*

$$0 \longrightarrow N \longrightarrow M \longrightarrow Q \longrightarrow 0.$$

If

$$\chi_{\mathfrak{m}}(N, T) = e_d(N)T^d + \text{lower terms}$$

and

$$\chi_{\mathfrak{m}}(Q, T) = e_d(Q)T^d + \text{lower terms}$$

then

$$\chi_{\mathfrak{m}}(M, T) = (e_d(N) + e_d(Q))T^d + \text{lower terms}.$$

Proof. Consider the exact sequence

$$0 \longrightarrow N/N \cap \mathfrak{m}^i M \longrightarrow M/\mathfrak{m}^i M \longrightarrow Q/\mathfrak{m}^i Q \longrightarrow 0.$$

From this follows

$$\ell_A(N/N \cap \mathfrak{m}^i M) = \chi_{\mathfrak{m}}(M, i) - \chi_{\mathfrak{m}}(Q, i), \qquad i \gg 0.$$

By the Artin–Rees lemma we can find $k \in \mathbb{N}$, such that

$$N \cap \mathfrak{m}^i M = \mathfrak{m}^{i-k}(N \cap \mathfrak{m}^k M), \qquad i \geq k$$

and whence

$$\mathfrak{m}^i N \subseteq N \cap \mathfrak{m}^i M \subseteq \mathfrak{m}^{i-k} N$$

and whence

$$\chi_{\mathfrak{m}}(N, i - k) \leq \ell_A(N/N \cap \mathfrak{m}^i M) \leq \chi_{\mathfrak{m}}(N, i), \qquad i \gg 0$$

and by the result above

$$\chi_{\mathfrak{m}}(N, i - k) \leq \chi_{\mathfrak{m}}(M, i) - \chi_{\mathfrak{m}}(Q, i) \leq \chi_{\mathfrak{m}}(N, i), \qquad i \gg 0.$$

The result follows from these inequalities by an elementary consideration. □

Definition 1.40. The supremum of $r \in \mathbb{N}$, for which there exists a chain of prime ideals in A of length r, is called the *dimension* of the local ring A, and denoted $\dim(A)$.

A sequence of length $\dim(A)$ of elements in $\mathfrak{m}$ generating an ideal of finite colength is called a *system of parameters*.

Theorem 1.41 (Dimension theorem). *For a noetherian local ring A the following integers are the same*

(1) $\dim(A)$.

(2) $\deg(\chi_m(A, T))$.

(3) *The least number of elements in $\mathfrak{m}$ needed to generate an ideal of finite colength.*

Proof. Let d_1, d_2, d_3 denote the integers defined in (1), (2), (3) respectively. We have $d_1 \leq d_2$ by 1.38 above and $d_2 \leq d_3$ by 1.42 below. We shall prove that $d_3 \leq d_1$ by induction on $\dim(A)$. If $\dim(A) = 0$, then A is an artinian local ring, 1.18, i.e., the ideal 0 has finite colength. Suppose $\dim(A) > 0$. Let $\mathfrak{q}_1, \ldots, \mathfrak{q}_e$ denote the minimal prime ideals in A, 1.14. It follows from 1.5 that $\mathfrak{m} \neq \bigcup \mathfrak{q}_i$. Pick $a \in \mathfrak{m}$ which is not contained in any of the $\mathfrak{q}_i$'s. Clearly $\dim(A/(a)) < \dim(A)$, thus by the induction hypothesis we can find $a_2, \ldots, a_{d_1} \in \mathfrak{m}$ whose residue classes $\mod a$ generate an ideal of finite colength. It follows that $a, a_2, \ldots, a_{d_1}$ generate an ideal of finite colength. $\square$

Lemma 1.42. *Let $a_1, \ldots, a_s \in \mathfrak{m}$ generate an ideal of finite colength. For a finitely generated module M, we have*

$$s \geq \deg(\chi_\mathfrak{m}(M, T)).$$

Proof. Let $\mathfrak{q} = (a_1, \ldots, a_s)$. By the proof of 1.36 we can find $\chi_\mathfrak{q}(M, T) \in \mathbb{Q}[T]$ of degree $\leq s$ such that

$$\ell_A(M/\mathfrak{q}^i M) = \chi_\mathfrak{q}(M, i), \qquad i \gg 0.$$

On the other hand, we have $\mathfrak{m}^k \subseteq \mathfrak{q} \subseteq \mathfrak{m}$ for some $k \in \mathbb{N}$. From this follows easily, that

$$\deg(\chi_\mathfrak{m}(M, T)) = \deg(\chi_\mathfrak{q}(M, T)). \qquad \square$$

Corollary 1.43 (Krull's principal ideal theorem). *Let $a \in A$ be an element of the maximal ideal and let $\mathfrak{q}$ be a prime ideal containing a. If $\mathfrak{q}$ is a minimal prime ideal in $A/(a)$ then any prime ideal in A properly contained in $\mathfrak{q}$ is minimal.*

Proof. Consider $A_\mathfrak{q}$. $\square$

Corollary 1.44. *Let $a \in A$ be an element of the maximal ideal. Then*

$$\dim(A/(a)) \geq \dim(A) - 1.$$

Proof. Put $d - 1 = \dim(A/(a))$ and choose $a_2, \ldots, a_d \in \mathfrak{m}$ such that the residue classes of these elements mod a generates an ideal of finite colength. Then $d \geq \dim(A)$ by 1.42. $\qquad\square$

Corollary 1.45. *Let $a \in \mathfrak{m}$ be a nonzero divisor in A. Then*

$$\dim(A/(a)) = \dim(A) - 1.$$

Proof. Follows from 1.44 and 1.39 applied to the exact sequence

$$0 \longrightarrow A \xrightarrow{\ a\ } A \longrightarrow A/(a) \longrightarrow 0. \qquad\square$$

1.9 Regular local rings

Throughout this section A denotes a noetherian local ring with maximal ideal $\mathfrak{m}$ and residue field k.

Example 1.46. It follows from the Dimension theorem, 1.41 and Nakayama's lemma, 1.3 that

$$\dim(A) \leq \operatorname{rank}_k(\mathfrak{m}/\mathfrak{m}^2).$$

Definition 1.47. The local ring A is called *regular* if

$$\dim(A) = \operatorname{rank}_k(\mathfrak{m}/\mathfrak{m}^2).$$

Proposition 1.48. *The noetherian local ring A is regular if and only if the map*

$$k[X_1, \ldots, X_d] \to \operatorname{gr}_{\mathfrak{m}}(A), \qquad X_i \mapsto x_i$$

is an isomorphism for a basis $x_1, \ldots x_d$ of the k-vectorspace $\mathfrak{m}/\mathfrak{m}^2$.

Proof. Suppose $k[X_1, \ldots, X_d] \to \operatorname{gr}_{\mathfrak{m}}(A)$ is an isomorphism. Then by 1.37 and 1.29

$$\Delta\chi_{\mathfrak{m}}(A, T) = \mathfrak{H}(k[X_1, \ldots, X_d], T) = \binom{T + d - 1}{d - 1},$$

from which we may conclude that $\deg(\chi_{\mathfrak{m}}(A, T)) = d$, that is, $\dim(A) = \operatorname{rank}_k(\mathfrak{m}/\mathfrak{m}^2)$. Conversely, if $\deg(\chi_{\mathfrak{m}}(A, T)) = d$, then $\deg(\mathfrak{H}(\operatorname{gr}_{\mathfrak{m}}(A), T)) = d - 1$ which implies that $k[X_1, \ldots, X_d] \to \operatorname{gr}_{\mathfrak{m}}(A)$ is an isomorphism, as it follows from 1.30. $\qquad\square$

Corollary 1.49. *A regular local ring is an integral domain.*

Proof. We shall prove more generally, that if $\mathrm{gr}_{\mathfrak{m}}(A)$ is an integral domain, then A itself is an integral domain. Recall from Krull intersection theorem, 1.35, that $\bigcap_i \mathfrak{m}^i = 0$. For $a \in A - \{0\}$, let $\nu(a) \in \mathbb{N}$ be such that $a \in \mathfrak{m}^{\nu(a)} - \mathfrak{m}^{\nu(a)+1}$. The residue class of $a \mod \mathfrak{m}^{\nu(a)+1}$ will be denoted $\bar{a} \in \mathfrak{m}^{\nu(a)}/\mathfrak{m}^{\nu(a)+1}$. Let now $a, b \in A - 0$, we have $\bar{a}\bar{b} \neq 0$, i.e., $ab \notin \mathfrak{m}^{\nu(a)+\nu(b)+1}$, in particular $ab \neq 0$. $\qquad\square$

Definition 1.50. Suppose A is a regular local ring of dimension d. A *regular system of parameters* for A is a sequence $x_1, \ldots, x_d \in \mathfrak{m}$, whose residue classes $\mod \mathfrak{m}^2$ is a basis for the k-vector space $\mathfrak{m}/\mathfrak{m}^2$.

Proposition 1.51. *Let A be a regular local ring and $\mathfrak{p} \subset \mathfrak{m}$ an ideal of A. Then $A/\mathfrak{p}$ is a regular local ring if and only if $\mathfrak{p}$ is generated by a subset of a regular system of parameters for A.*

Proof. Let $x_1, \ldots, x_d$ denote a regular system of parameters. Let us show that $A/(x_1)$ is a regular local ring. Let $\mathfrak{m}_1$ denote its maximal ideal. Clearly

$$\mathrm{rank}_k(\mathfrak{m}_1/\mathfrak{m}_1^2) = \mathrm{rank}_k(\mathfrak{m}/\mathfrak{m}^2) - 1.$$

On the other hand, since A is an integral domain and $x_1 \neq 0$, we have by 1.45

$$\dim(A/(x_1)) = \dim(A) - 1.$$

which proves that $A/(x_1)$ is regular. It is now clear that the residue classes of $x_2, \ldots, x_d \mod x_1$ is a regular system of parameters. It now follows by induction, that if $\mathfrak{p}$ is generated by a subset of a regular system of parameters, then $A/\mathfrak{p}$ is regular.

Conversely, suppose $A/\mathfrak{p}$ is regular of dimension $d - e$. Let $\mathfrak{n}$ denote the maximal ideal in $A/\mathfrak{p}$ and note that we have an exact sequence

$$\mathfrak{p} \longrightarrow \mathfrak{m}/\mathfrak{m}^2 \longrightarrow \mathfrak{n}/\mathfrak{n}^2 \longrightarrow 0,$$

whence we can choose a regular system of parameters $x_{e+1}, \ldots, x_d$ for A such that the residue classes $\mod \mathfrak{p}$ of $x_{e+1}, \ldots, x_d$ form a regular system of parameters for $A/\mathfrak{p}$, and such that $x_1, \ldots, x_e \in \mathfrak{p}$. We are going to prove that $\mathfrak{p} = (x_1, \ldots, x_e)$. By our previous result $A/(x_1, \ldots, x_e)$ is a regular local ring of dimension $d - e$. The two ideals $(x_1, \ldots, x_e) \subseteq \mathfrak{p}$ are both prime ideals, 1.49, and the residue rings of A with respect to both have the same dimension. Then they must be equal. $\qquad\square$

Chapter 2

Modules over a Local Ring

2.1 Support of a module

Let A be a ring, then $\operatorname{Spec}(A)$ denotes the set of prime ideals in A. For an ideal $\mathfrak{a}$ in A, we let $V(\mathfrak{a})$ denote the set of prime ideals containing $\mathfrak{a}$. $V(\mathfrak{a})$ may be identified with $\operatorname{Spec}(A/\mathfrak{a})$. For a prime ideal $\mathfrak{p}$, $\operatorname{Spec}(A_{\mathfrak{p}})$ may be identified with the subset of prime ideals contained in $\mathfrak{p}$. Subsets of $\operatorname{Spec}(A)$ of the form $V(\mathfrak{a})$, $\mathfrak{a}$ an ideal, satisfies the axioms for closed sets in a topological space. Equipped with this topology, the *Zariski topology*, $\operatorname{Spec}(A)$ is called the *spectrum* of A.

Definition 2.1. Let M be an A-module, define the *support* of M

$$\operatorname{Supp}(M) = \{\mathfrak{p} \in \operatorname{Spec}(A) \mid M_{\mathfrak{p}} \neq 0\}.$$

Proposition 2.2. *Suppose M is a finitely generated A-module then*

$$\operatorname{Supp}(M) = V(\operatorname{Ann}(M)).$$

Proof. Straightforward. $\qquad\square$

Proposition 2.3. *Suppose M and N are finitely generated A-modules, then*

$$\operatorname{Supp}(M \otimes_A N) = \operatorname{Supp}(M) \cap \operatorname{Supp}(N).$$

Proof. Straightforward from Nakayama's lemma, 1.3. $\qquad\square$

Corollary 2.4. *Let A be a ring, $\mathfrak{a}$ an ideal in A and M a finitely generated A-module. Then*

$$\operatorname{Supp}(M/\mathfrak{a}M) = \operatorname{Supp}(M) \cap V(\mathfrak{a}).$$

Proof. Straightforward from Proposition 2.3. $\qquad\square$

Proposition 2.5. *Let N be a submodule of the module M. Then* $\operatorname{Supp}(N) \subseteq \operatorname{Supp}(M)$ *and*

$$\operatorname{Supp}(M) = \operatorname{Supp}(N) \cup \operatorname{Supp}(M/N).$$

Proof. Straightforward.   □

2.2   Associated prime ideals

Throughout this section A denotes a noetherian ring.

Definition 2.6. Given an A-module M. A prime ideal $\mathfrak{p}$ in A is said to be *associated* to M if there exists $x \in M$ whose annihilator $\operatorname{Ann}(x) = \{a \in A \mid ax = 0\}$ equals $\mathfrak{p}$. The set of prime ideals associated to M is denoted $\operatorname{Ass}(M)$.

Proposition 2.7. *Let M be an A-module. Then $M = 0$ if and only if* $\operatorname{Ass}(M) = \emptyset$.

Proof. Suppose $M \neq 0$. As in the proof of 1.13 a maximal annihilator different from A is a prime ideal.   □

Example 2.8. Let $\mathfrak{p}$ be a prime ideal in A. Then

$$\operatorname{Ass}(A/\mathfrak{p}) = \{\mathfrak{p}\}.$$

Proposition 2.9. *Let N be a submodule of the module M. Then*

$$\operatorname{Ass}(N) \subseteq \operatorname{Ass}(M) \subseteq \operatorname{Ass}(N) \cup \operatorname{Ass}(M/N).$$

Proof. The first inclusion is trivial. As for the second, suppose given $\mathfrak{p} \in \operatorname{Ass}(M)$ such that $\mathfrak{p} \notin \operatorname{Ass}(N)$. Choose a submodule P of M such that $P \simeq A/\mathfrak{p}$. We have

$$\operatorname{Ass}(P \cap N) \subseteq \operatorname{Ass}(P) \cap \operatorname{Ass}(N)$$

from which it follows that $P \cap N = 0$, and therefore M/N contains a submodule isomorphic to $A/\mathfrak{p}$.   □

Lemma 2.10. *Let $M \neq 0$ be a finitely generated A-module and $\mathfrak{p} \in \operatorname{Ass}(M)$. Then there exists a submodule N of M with*

$$\operatorname{Ass}(N) = \{\mathfrak{p}\}, \quad \operatorname{Ass}(M/N) = \operatorname{Ass}(M) - \{\mathfrak{p}\}.$$

Proof. Choose N maximal in the set of submodules N' of M for which $\mathrm{Ass}(N') = \{\mathfrak{p}\}$. Let us prove $\mathrm{Ass}(M/N) = \mathrm{Ass}(M) - \{\mathfrak{p}\}$. Let $\mathfrak{q} \in \mathrm{Ass}(M/N)$ and choose a submodule Q of M containing N and with $Q/N \simeq A/\mathfrak{q}$. We have $\mathrm{Ass}(Q) \subseteq \{\mathfrak{p}, \mathfrak{q}\}$ and $\mathrm{Ass}(Q) \neq \{\mathfrak{p}\}$, whence $\mathfrak{q} \in \mathrm{Ass}(Q)$ and therefore $\mathfrak{q} \in \mathrm{Ass}(M)$. On the other hand, $\mathrm{Ass}(M) \subseteq \{\mathfrak{p}\} \cup \mathrm{Ass}(M/N)$ and $\mathfrak{p} \notin \mathrm{Ass}(M/N)$ since this would contradict maximality of N. $\qquad\square$

Proposition 2.11. *Let M be a finitely generated A-module, then $\mathrm{Ass}(M)$ is a finite set.*

Proof. Follows immediately from 2.9 and 1.13. $\qquad\square$

Corollary 2.12. *Let M be a finitely generated A-module. For each $\mathfrak{p} \in \mathrm{Ass}(M)$ there is a submodule $Q(\mathfrak{p}) \subseteq M$ such that $\mathrm{Ass}(Q(\mathfrak{p})) = \mathrm{Ass}(M) - \{\mathfrak{p}\}$ and $\mathrm{Ass}(M/Q(\mathfrak{p})) = \{\mathfrak{p}\}$. M injects*

$$0 \longrightarrow M \longrightarrow \bigoplus_{\mathfrak{p} \in \mathrm{Ass}(M)} M/Q(\mathfrak{p}).$$

Proof. By 2.11 the submodule $Q(\mathfrak{p})$ is constructed by finitely many applications of 2.10. $\mathrm{Ass}(\bigcap_{\mathfrak{p}} Q(\mathfrak{p})) = \emptyset$, so conclusion by 2.7. $\qquad\square$

Proposition 2.13. *Let S denote a multiplicative subset of A and M an A-module. Then*

$$\mathrm{Ass}_{S^{-1}A}(S^{-1}M) = \{\mathfrak{p} \in \mathrm{Ass}(M) \mid \mathfrak{p} \cap S = \emptyset\}.$$

Proof. Straightforward, left to the reader. $\qquad\square$

Proposition 2.14. *Let A be a noetherian ring and M a finitely generated A-module. Then*

$$\mathrm{Ass}(M) \subseteq \mathrm{Supp}(M)$$

and any $\mathfrak{p} \in \mathrm{Supp}(M)$, which is minimal in $\mathrm{Supp}(M)$, is contained in $\mathrm{Ass}(M)$.

Proof. The first assertion is obvious from the definition. Assume $\mathfrak{p} \in \mathrm{Supp}(M)$ is minimal. Then the $A_{\mathfrak{p}}$-module $M_{\mathfrak{p}}$ has support exactly in the maximal ideal, whence $\{\mathfrak{p}A_{\mathfrak{p}}\} = \mathrm{Ass}(M_{\mathfrak{p}})$. Conclusion by 2.13. $\qquad\square$

Proposition 2.15. *Let M and N be finitely generated A-modules. Then*

$$\mathrm{Ass}(\mathrm{Hom}_A(M, N)) = \mathrm{Supp}(M) \cap \mathrm{Ass}(N).$$

Proof. Let us first remark that if $\mathfrak{p}$ is a prime ideal, then the canonical map

$$\mathrm{Hom}_A(M, N)_\mathfrak{p} \to \mathrm{Hom}_{A_\mathfrak{p}}(M_\mathfrak{p}, N_\mathfrak{p})$$

is an isomorphism as it follows by considering a finite presentation of M. We want to show that

$$\mathfrak{p} \in \mathrm{Ass}(\mathrm{Hom}_A(M, N)) \quad \text{if and only if} \quad M_\mathfrak{p} \neq 0 \text{ and } \mathfrak{p} \in \mathrm{Ass}(N).$$

By the previous remark we may assume that A is local and that $\mathfrak{p}$ is the maximal ideal of A. Put $k = A/\mathfrak{p}$. We have

$$\begin{aligned}
\mathrm{Hom}_A(k, \mathrm{Hom}_A(M, N)) &= \mathrm{Hom}_A(M, \mathrm{Hom}_A(k, N)) \\
&= \mathrm{Hom}_k(M \otimes_A k, \mathrm{Hom}_A(k, N)).
\end{aligned}$$

Conclusion by Nakayama's lemma, 1.3. $\qquad\qquad\square$

Definition 2.16. Let M be an A-module. An element $a \in A$ is called a *nonzero divisor* for M, if scalar multiplication with a on M is an injective endomorphism.

Proposition 2.17. *Let M be an A-module and $a \in A$. Then a is a nonzero divisor on M if and only if*

$$a \notin \bigcup_{\mathfrak{p} \in \mathrm{Ass}(M)} \mathfrak{p}.$$

Proof. Suppose $a \notin \bigcup_{\mathfrak{p} \in \mathrm{Ass}(M)} \mathfrak{p}$. We want to show that the kernel $K = \mathrm{Ker}(a : M \to M)$ is zero, or what amounts to the same, that $\mathrm{Ass}(K) = \emptyset$. For $\mathfrak{q} \in \mathrm{Ass}(K)$ we have $a \in \mathfrak{q}$ and $\mathfrak{q} \in \mathrm{Ass}(M)$, whence $\mathrm{Ass}(K) = \emptyset$. $\quad\square$

2.3 Dimension of a module

Definition 2.18. Let $M \neq 0$ be a module over the noetherian local ring A. Define the *dimension*

$$\dim(M) = \sup_{\mathfrak{p} \in \mathrm{Ass}(M)} \dim(A/\mathfrak{p}).$$

Notice that

$$\dim(M) = \deg \chi_\mathfrak{m}(M, T).$$

Namely consider a filtration of M

$$0 = M_0 \subseteq M_1 \subseteq \cdots \subseteq M_r = M$$

such that $M_i/M_{i-1} \simeq A/\mathfrak{q}_i$, $i = 1, \ldots, r$, where $\mathfrak{q}_i$ is a prime ideal. It follows from 1.39 that

$$\deg \chi_\mathfrak{m}(M, T) = \sup_i \deg \chi_\mathfrak{m}(A/\mathfrak{q}_i, T).$$

Let us note that $\mathrm{Ass}(M)$ and $\{\mathfrak{q}_i \mid i = 1, \ldots, n\}$ have the same minimal elements. Whence

$$\deg \chi_\mathfrak{m}(M, T) = \sup_{\mathfrak{p} \in \mathrm{Ass}(M)} \deg \chi_\mathfrak{m}(A/\mathfrak{p}, T).$$

Conclusion by 1.41.

Theorem 2.19. *Let $0 \neq M$ be a finitely generated module over a noetherian local ring A and let a be contained in the maximal ideal. Then*

(1) $\dim(M/aM) \geq \dim(M) - 1$.

(2) $\dim(M/aM) = \dim(M) - 1$ *if and only if a is outside those prime ideals $\mathfrak{p} \in \mathrm{Ass}(M)$ for which $\dim(M) = \dim(A/\mathfrak{p})$.*

Proof. Notice by 2.4 that

$$\begin{aligned}
\mathrm{Supp}(M/aM) &= \mathrm{Supp}(M) \cap V(a) \\
&= V(\mathrm{Ann}(M)) \cap V(a) \\
&= V((\mathrm{Ann}(M), a)).
\end{aligned}$$

Put $B = A/\mathrm{Ann}(M)$ and let b denote the residue class of a. Then $\dim(M) = \dim(B)$, $\dim(M/aM) = \dim(B/(b))$ and the minimal elements in $\mathrm{Ass}(M)$ correspond to the minimal prime ideals in B. Conclusion by 1.44 and the Dimension theorem, 1.41, giving the dimension by means of chains of prime ideals. □

Corollary 2.20. *Let A be a noetherian local ring, M a finitely generated A-module and a contained in the maximal ideal a nonzero divisor for M. Then*

$$\dim(M/aM) = \dim(M) - 1.$$

Proof. Follows immediately from 2.19 and 1.45. □

2.4   Depth of a module

Throughout this section A denotes a noetherian local ring with maximal ideal $\mathfrak{m}$ and residue field k.

Lemma 2.21. *Let $M \neq 0$ be a finitely generated A-module. Then there exists $i \in \mathbb{N}$ such that*

$$\mathrm{Ext}_A^i(k, M) \neq 0.$$

Proof. Suppose $\mathfrak{m} \in \mathrm{Ass}(M)$ and notice that

$$\mathrm{Hom}_A(k, M) = \{x \in M \mid \mathfrak{m}x = 0\},$$

whence $\mathrm{Hom}_A(k, M) \neq 0$ in this case. Let us now prove the lemma by induction on $\dim(M)$, the case $\dim(M) = 0$ being already taken care of. We may also assume $\mathfrak{m} \notin \mathrm{Ass}(M)$. By 1.5 and 2.11 it follows that $\mathfrak{m} - \bigcup_{\mathfrak{p} \in \mathrm{Ass}(M)} \mathfrak{p}$ is nonempty thus we can find $a \in \mathfrak{m}$ a nonzero divisor for M. The long Ext-sequence arising from the short exact sequence

$$0 \longrightarrow M \overset{a}{\longrightarrow} M \longrightarrow M/aM \longrightarrow 0$$

yields

$$\mathrm{Ext}_A^{i-1}(k, M) \longrightarrow \mathrm{Ext}_A^{i-1}(k, M/aM) \longrightarrow \mathrm{Ext}_A^i(k, M)$$

from which the conclusion follows.   □

Definition 2.22. Let A be a noetherian local ring, $M \neq 0$ a finitely generated A-module. Define the *depth* of M by

$$\mathrm{depth}(M) = \inf\{i \in \mathbb{N} \mid \mathrm{Ext}_A^i(k, M) \neq 0\}.$$

Proposition 2.23. *Let $M \neq 0$ be a finitely generated A-module and $a \in \mathfrak{m}$ a nonzero divisor for M. Then*

$$\mathrm{depth}(M/aM) = \mathrm{depth}(M) - 1.$$

Proof. Follows immediately from the long exact sequence written in the proof of 2.21.   □

Definition 2.24. Let A be a noetherian local ring, $M \neq 0$ a finitely generated A-module. A sequence $a_1, \ldots, a_r$ of elements in the maximal ideal is called a *M-regular sequence* of length r if for all $i = 1, \ldots, r$, a_i is a nonzero divisor on $M/(a_1, \ldots, a_{i-1})M$.

Proposition 2.25. *Let $M \neq 0$ be a finitely generated A-module. Then* $\mathrm{depth}(M)$ *is the maximal length of any M-regular sequence.*

Proof. If $\mathrm{depth}(M) > 0$ there is a nonzero divisor on M, 2.15. The conclusion follows by induction and 2.23. □

Theorem 2.26. *Let $M \neq 0$ be a finitely generated A-module and $\mathfrak{p} \in \mathrm{Ass}(M)$. Then*

$$\mathrm{depth}(M) \leq \dim(A/\mathfrak{p}) \leq \dim(M).$$

Proof. Only the inequality to the left has to be proven. Induction on $\mathrm{depth}(M)$. The case $\mathrm{depth}(M) = 0$ being trivial. Suppose $\mathrm{depth}(M) > 0$. As in the proof of 2.21 we may choose $a \in A$ a nonzero divisor for M. Choose a minimal element $\mathfrak{q}$ in $V(\mathfrak{p}) \cap V(a)$. We are going to prove that $\mathfrak{q} \in \mathrm{Ass}(M/aM)$. For this we choose a submodule N of M with $\mathrm{Ass}(N) = \{\mathfrak{p}\}$ and $\mathrm{Ass}(M/N) \subseteq \mathrm{Ass}(M)$ as can be done by the 2.10. We have

$$\mathrm{Supp}(N/aN) = \mathrm{Supp}(N) \cap V(a)$$
$$= V(\mathfrak{p}) \cap V(a).$$

Thus $\mathfrak{q} \in \mathrm{Ass}(N/aN)$ as it follows from 2.14. Consider the exact commutative diagram:

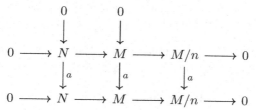

Since $\mathrm{Ass}(M/N) \subseteq \mathrm{Ass}(M)$, $a : M/N \to M/N$ is injective, and it follows by the snake lemma that $N/aN \to M/aM$ is injective, thus $\mathfrak{q} \in \mathrm{Ass}(M/aM)$. □

2.5 Cohen–Macaulay modules

Throughout this section A denotes a noetherian local ring with maximal ideal $\mathfrak{m}$ and residue field k.

Definition 2.27. A finitely generated module $M \neq 0$ is called a *Cohen–Macaulay* module if

$$\text{depth}(M) = \dim(M).$$

The local ring A is called a *Cohen–Macaulay* ring if the A-module A is a Cohen–Macaulay module.

Example 2.28. A regular local ring, 1.9 is a Cohen–Macaulay ring.

Theorem 2.29. *Let $M \neq 0$ be a finitely generated Cohen–Macaulay module and $a_1, \ldots, a_r \in \mathfrak{m}$ such that*

$$\dim(M/(a_1, \ldots, a_r)M) = \dim(M) - r.$$

Then $M/(a_1, \ldots, a_r)M$ is a Cohen–Macaulay module and $a_1, \ldots, a_r$ is a M-regular sequence.

Proof. By 2.19 we have

$$\dim(M/a_1 M) = \dim(M) - 1$$

and for $N = M/a_1 M$

$$\dim(N/(a_2, \ldots, a_r)N) = \dim(N) - (r - 1).$$

Thus by simple induction consideration it suffices to treat the case $r = 1$. By 2.19 we have $a_1 \notin \mathfrak{p}$, $\mathfrak{p} \in \text{Ass}(M)$, whence a_1 is a nonzero divisor for M. It follows from 2.20 and 2.23 that $M/a_1 M$ is a Cohen–Macaulay module. $\square$

Proposition 2.30. *Let $M \neq 0$ be a finitely generated Cohen–Macaulay module and $\mathfrak{p} \in \text{Supp}(M)$. Then $M_\mathfrak{p}$ is a Cohen–Macaulay $A_\mathfrak{p}$-module and*

$$\dim(A/\mathfrak{p}) + \dim_{A_\mathfrak{p}}(M_\mathfrak{p}) = \dim(M).$$

Proof. Put $r = \dim_{A_\mathfrak{p}}(M_\mathfrak{p})$ and choose by induction a sequence $a_1, \ldots, a_r \in \mathfrak{p}$ where a_i is a nonzero divisor for $M/(a_1, \ldots, a_{i-1})M$, $i = 1, \ldots, r$. We have $\dim((M/(a_1, \ldots, a_r)M)_\mathfrak{p}) = 0$, thus $\mathfrak{p} \in \text{Ass}(M/(a_1, \ldots, a_r)M)$ and therefore by 2.26 and 2.29

$$\dim(A/\mathfrak{p}) = \dim(M) - r.$$

Since $\dim((M/(a_1, \ldots, a_r)M)_\mathfrak{p}) = 0$ we get $\text{depth}((M/(a_1, \ldots, a_r)M)_\mathfrak{p}) = 0$ and whence

$$\text{depth}(M_\mathfrak{p}) = \dim(M_\mathfrak{p}). \qquad \square$$

Theorem 2.31. *Suppose there exists a finitely generated Cohen–Macaulay module M with $\mathrm{Supp}(M) = \mathrm{Spec}(A)$. Then for any two prime ideals $\mathfrak{p} \subseteq \mathfrak{q}$ we have*

$$\dim(A/\mathfrak{p}) = \dim(A_\mathfrak{q}/\mathfrak{p}_\mathfrak{q}) + \dim(A/\mathfrak{q}).$$

Proof. It follows from the proof of 2.30 that we can find a Cohen–Macaulay module N with $\mathfrak{p} \in \mathrm{Ass}(N)$. The formula in 2.30 yields

$$\dim(A/\mathfrak{q}) + \dim_{A_\mathfrak{p}}(N_\mathfrak{p}) = \dim(N).$$

Since N and $N_\mathfrak{q}$ are Cohen–Macaulay modules with $\mathfrak{p} \in \mathrm{Ass}(N)$ and $\mathfrak{p}_\mathfrak{q} \in \mathrm{Ass}(N_\mathfrak{q})$ we get

$$\dim(A/\mathfrak{p}) = \dim(N) \quad \text{and} \quad \dim(A_\mathfrak{q}/\mathfrak{p}_\mathfrak{q}) = \dim(N_\mathfrak{q}). \qquad \square$$

Remark 2.32. A noetherian ring is called *catenary* if any two saturated chains of prime ideals with the same endpoints have the same length. It is easily seen that the conclusion of 2.31 is that a ring is catenary if there exists a Cohen–Macaulay module which is supported in all prime ideals in the ring. Not all rings are catenary.

2.6   Modules of finite projective dimension

Let A denote a ring, an A-module P is called *projective* if the functor $\mathrm{Hom}_A(P, -)$ is exact. If the functor $- \otimes_A P$ is exact, P is called *flat*. A projective module is flat and a finitely generated flat module over a noetherian ring is projective. The smallest $d \in \mathbb{N}$ for which M has a resolution of length d by projective modules is called the *projective dimension* and is denoted $\mathrm{proj\,dim}(M)$.

Throughout the rest of this section A denotes a noetherian local ring with maximal ideal $\mathfrak{m}$ and residue field k.

Lemma 2.33. *Let M be a finitely generated A-module. If*

$$\mathrm{Tor}_1^A(k, M) = 0,$$

then M is a free module. A finitely generated flat module is free.

Proof. Choose a free module L and a linear map $\phi : L \to M$ such that $1 \otimes \phi : k \otimes_A L \to k \otimes_A M$ is an isomorphism. Note that $k \otimes_A \mathrm{Cok}(\phi) = 0$ and whence $\mathrm{Cok}(\phi) = 0$ by Nakayama's lemma, 1.3. Consider the exact sequence

$$0 \longrightarrow \mathrm{Tor}_1^A(k, M) \longrightarrow k \otimes_A \mathrm{Ker}(\phi) \longrightarrow k \otimes_A L \longrightarrow k \otimes_A M \longrightarrow 0$$

to see that $k \otimes_A \mathrm{Ker}(\phi) = 0$ and whence $\mathrm{Ker}(\phi) = 0$. $\qquad\square$

Lemma 2.34. *Let* $\phi : M \to L$ *be a linear map between finitely generated A-modules and suppose L is a free module. If*

$$1 \otimes \phi : k \otimes_A M \to k \otimes_A L$$

is injective, then ϕ has a retraction $\psi : L \to M$, $\psi \circ \phi = \mathbf{1}_M$.

Proof. The factorization of ϕ

$$M \to \mathrm{Im}(\phi) \to L$$

shows that

$$k \otimes_A M \to k \otimes_A \mathrm{Im}(\phi), \qquad \text{is bijective}$$

and

$$k \otimes_A \mathrm{Im}(\phi) \to k \otimes_A L, \qquad \text{is injective.}$$

The exact sequence

$$0 \longrightarrow \mathrm{Im}(\phi) \longrightarrow L \longrightarrow \mathrm{Cok}(\phi) \longrightarrow 0$$

gives rise to the exact sequence

$$0 \longrightarrow \mathrm{Tor}_1^A(k, \mathrm{Cok}(\phi)) \longrightarrow k \otimes_A \mathrm{Im}(\phi) \longrightarrow k \otimes_A L,$$

which shows that $\mathrm{Tor}_1^A(k, \mathrm{Cok}(\phi)) = 0$ and therefore by 2.33 that $\mathrm{Cok}(\phi)$ is a free module. This implies that $\mathrm{Im}(\phi)$ is a free module, with the consequence that

$$0 \longrightarrow \mathrm{Ker}(\phi) \longrightarrow M \longrightarrow \mathrm{Im}(\phi) \longrightarrow 0$$

is split exact. Consider the exact sequence

$$0 \longrightarrow k \otimes_A \mathrm{Ker}(\phi) \longrightarrow k \otimes_A M \longrightarrow k \otimes_A \mathrm{Im}(\phi) \longrightarrow 0$$

and recall that $k \otimes_A M \to k \otimes_A \mathrm{Im}(\phi)$ is injective, to see that $\mathrm{Ker}(\phi) = 0$ by Nakayama's lemma, 1.3. $\qquad\square$

Proposition 2.35. *Let M be a finitely generated A-module with*

$$\operatorname{Tor}_{d+1}^{A}(k, M) = 0.$$

Then M has a resolution of length at most d by finitely generated free modules.

Proof. Consider a resolution of M by finitely generated free modules

$$L_{d+2} \xrightarrow{\partial_{d+2}} L_{d+1} \xrightarrow{\partial_{d+1}} L_d \cdots L_1 \xrightarrow{\partial_1} L_0 \longrightarrow M \longrightarrow 0.$$

We have an exact sequence

$$0 \longrightarrow H_{d+1}(k \otimes_A L.) \longrightarrow k \otimes_A \operatorname{Cok}(\partial_{d+2}) \longrightarrow k \otimes_A L_d.$$

By assumption $H_{d+1}(k \otimes_A L.) = 0$. Thus by 2.34 the map $\operatorname{Cok}(\partial_{d+2}) \to L_d$ has a retraction, i.e., $\operatorname{Im}(\partial_{d+1})$ is a direct summand in L_d. $\qquad\square$

Remark 2.36. Let M be a finitely generated A-module. The smallest $d \in \mathbb{N}$ for which M has a resolution of length d by finitely generated free modules is the projective dimension $\operatorname{proj dim}(M)$.

Remark 2.37. Let M be a finitely generated A-module. A resolution $L. \to M$ by finitely generated free modules is *minimal* if all the differentials $\partial_n \otimes 1 : L_n \otimes_A k \to L_{n-1} \otimes_A k$ are zero. By Nakayama's lemma, 1.3, minimal resolutions exist and are unique up to isomorphism of complexes.

2.7 The Koszul complex

Let A denote a ring, for $a \in A$ let $K.(a)$ denote the complex

$$K_i(a) = \begin{cases} A, & i = 0, 1 \\ 0, & i \neq 0, 1, \end{cases}$$

the differential being scalar multiplication with a.

Recall that if $K.$ and $Q.$ are complexes we define the *tensor complex* $K. \otimes_A Q.$ by

$$(K. \otimes_A Q.)_n = \bigoplus_{i \in \mathbb{Z}} K_i \otimes_A Q_{n-i},$$

$$\partial_n = \partial_i \otimes 1 + (-1)^i 1 \otimes \partial_{n-i}.$$

Definition 2.38. Let $a. = (a_1, \ldots, a_r)$ be a sequence of elements in A. The complex

$$K.(a.) = K.(a_1) \otimes_A \cdots \otimes_A K.(a_r)$$

is called the *Koszul complex* relative to $a.$.

Theorem 2.39. *Let M be an A-module and suppose that $a. = (a_1, \ldots, a_r)$ is a sequence of elements in A such that for all $i = 1, \ldots, r$, the element a_i is a nonzero divisor on $M/(a_1, \ldots, a_{i-1})M$. Then*

$$H_i(K.(a.) \otimes_A M) = \begin{cases} M/(a_1, \ldots, a_r)M, & i = 0 \\ 0, & i \neq 0. \end{cases}$$

Proof. Follows by induction and the following 2.40. $\qquad\square$

Lemma 2.40. *Let $Q.$ and $K.$ be complexes of A-modules. Suppose that $K.$ is a complex of projective modules and that $K_i = 0$ for $i \neq 0, 1$. Then for each $p \in \mathbb{Z}$ there is an exact sequence*

$$0 \longrightarrow H_0(K. \otimes_A H_p(Q.)) \longrightarrow H_p(K. \otimes_A Q.)$$
$$\longrightarrow H_1(K. \otimes_A H_{p-1}(Q.)) \longrightarrow 0.$$

Proof. Exhibit an exact sequence of complexes

$$0 \longrightarrow K_0 \otimes_A Q. \longrightarrow K. \otimes_A Q. \longrightarrow K_1 \otimes_A Q.[-1] \longrightarrow 0$$

which on homology induces the exact sequence

$$K_1 \otimes_A H_p(Q.) \longrightarrow K_0 \otimes_A H_p(Q.) \longrightarrow H_p(K. \otimes_A Q.)$$
$$\longrightarrow K_1 \otimes_A H_{p-1}(Q.) \longrightarrow K_0 \otimes_A H_{p-1}(Q.). \qquad\square$$

Proposition 2.41. *Let A be a noetherian local ring and suppose that $a. = (a_1, \ldots, a_r)$ is a sequence of elements in the maximal ideal of A. Let M be a finitely generated A-module.*

  (i) *If $H_i(K.(a.) \otimes_A M) = 0$ then $H_j(K.(a.) \otimes_A M) = 0$ for $j \geq i$.*

  (ii) *If $H_1(K.(a.) \otimes_A M) = 0$ then $a_1, \ldots, a_r$ is a M-regular sequence.*

 (iii) *$(a_1, \ldots, a_r) + \mathrm{Ann}(M) \subseteq \mathrm{Ann}(H_i(K.(a.) \otimes_A M))$.*

Proof. Follows by Nakayama's lemma, 1.3, and 2.40. For (iii) remark that multiplication with a_1 on $K.(a_1)$ is homotopic to zero. $\qquad\square$

2.8 Regular local rings

Throughout this section A denotes a noetherian local ring with maximal ideal $\mathfrak{m}$ and residue field k.

Theorem 2.42. *If A is a regular local ring of dimension d, then any finitely generated module M has a resolution of length at most d by finitely generated free A-modules.*

Proof. Choose a regular system of parameters $a_1, \ldots, a_d$ for A. The Koszul complex $K_{\bullet}(a_1, \ldots a_d)$ yields a free resolution of k of length d, 1.51 and 2.39. Consequently $\mathrm{Tor}^A_{d+1}(k, M) = 0$, conclusion by 2.35.   □

Theorem 2.43. *If the A-module k has finite projective dimension then A is a regular local ring.*

Proof. Put $d = \mathrm{rank}_k(\mathfrak{m}/\mathfrak{m}^2)$. We are going to prove that there exists generators $a_1, \ldots, a_d$ for $\mathfrak{m}$ such that for each $i = 1, \ldots, d$, a_i is a nonzero divisor for $A/(a_1, \ldots, a_{i-1})$. Once this is done, we can conclude by 1.45.

We shall assume $d \geq 1$ and first prove that $\mathfrak{m} \notin \mathrm{Ass}(A)$. Choose $r \in \mathbb{N}$ such that $\mathrm{Tor}^A_{r+1}(k, k) = 0$, $\mathrm{Tor}^A_r(k, k) \neq 0$. Suppose $\mathfrak{m} \in \mathrm{Ass}(A)$. This implies the existence of a short exact sequence of the form

$$0 \longrightarrow k \longrightarrow A \longrightarrow E \longrightarrow 0.$$

The long exact sequence in $\mathrm{Tor}^A_{\bullet}(k, -)$ gives an isomorphism

$$\mathrm{Tor}^A_{r+1}(k, E) \simeq \mathrm{Tor}^A_r(k, k).$$

But $\mathrm{Tor}^A_{r+1}(k, k) = 0$ implies that k has a free resolution of length r and whence $\mathrm{Tor}^A_{r+1}(k, E) = 0$, giving a contradiction to the choice of r. Next let us show that there exists $a \in \mathfrak{m} - \mathfrak{m}^2$ which is a nonzero divisor in A. Let $\mathfrak{q}_1, \ldots, \mathfrak{q}_s$ denote the elements of $\mathrm{Ass}(A)$. We must show that, 2.17,

$$\mathfrak{q}_1 \cup \cdots \cup \mathfrak{q}_s \cup \mathfrak{m}^2 \neq \mathfrak{m}.$$

If these two expressions were equal we would have $\mathfrak{m} = \mathfrak{m}^2$ or $\mathfrak{m} \in \mathrm{Ass}(A)$ as it follows from 1.5. It follows from the following two lemmas that k has finite projective dimension as an $A/(a)$-module. Thus we can conclude the proof by induction on d.   □

Lemma 2.44. *Let $a \notin \mathfrak{m} - \mathfrak{m}^2$. Then we have a non canonical isomorphism*

$$\mathfrak{m}/a\mathfrak{m} \simeq k \oplus \mathfrak{m}/(a).$$

Proof. Pick elements $a_2, \ldots, a_d \in \mathfrak{m}$ such that $a, a_2, \ldots, a_d$ forms a basis for $\mathfrak{m}/\mathfrak{m}^2$. Note that

$$((a_2, \ldots, a_d) + a\mathfrak{m}) \cap (a) = a\mathfrak{m}.$$

This proves that the images of $(a_2, \ldots, a_d)$ and (a) in $\mathfrak{m}/a\mathfrak{m}$ form a direct sum decomposition of that module. Note $(a)/a\mathfrak{m} \simeq k$ by Nakayama's lemma, 1.3, and note that the composite

$$(a_2, \ldots, a_d)/(a_2, \ldots, a_d) \cap (a) \to \mathfrak{m}/a\mathfrak{m} \to \mathfrak{m}/(a)$$

is an isomorphism. $\qquad \square$

Lemma 2.45. *Let $a \in \mathfrak{m}$ be a nonzero divisor in A. Then for any A-module M for which a is a nonzero divisor, we have*

$$\mathrm{Tor}_i^{A/(a)}(k, M/aM) \simeq \mathrm{Tor}_i^A(k, M).$$

Proof. Use the exact sequence

$$0 \longrightarrow A \overset{a}{\longrightarrow} A \longrightarrow A/(a) \longrightarrow 0$$

to show that $\mathrm{Tor}_i^A(A/(a), M) = 0$ for $i \geq 1$. Let now $L. \to M$ be a free resolution of M. It follows that $L. \otimes_A A/(a)$ is a free resolution of the $A/(a)$-module M/aM. But

$$L. \otimes_A A/(a) \otimes_{A/(a)} k \simeq L. \otimes_A k. \qquad \square$$

Corollary 2.46. *Let $\mathfrak{p}$ be a prime ideal in the regular local ring A. Then $A_\mathfrak{p}$ is a regular local ring.*

Proof. A free resolution of $A/\mathfrak{p}$ gives a free resolution of $A_\mathfrak{p}/\mathfrak{p}A_\mathfrak{p}$. $\qquad \square$

2.9   Projective dimension and depth

Throughout this section A denotes a noetherian local ring with maximal ideal $\mathfrak{m}$ and residue field k.

Theorem 2.47 (Auslander–Buchsbaum's formula). *Let $M \neq 0$ be a finitely generated A-module of finite projective dimension. Then*

$$\mathrm{proj}\,\dim(M) + \mathrm{depth}(M) = \mathrm{depth}(A).$$

Proof. Induction on $\operatorname{proj dim}(M)$. The case $\operatorname{proj dim}(M) = 0$ is clear, let us next treat the case $\operatorname{proj dim}(M) = 1$. Choose a free module L and a linear map $\phi : L \to M$ such that $1 \otimes \phi : k \otimes_A L \to k \otimes_A M$ is an isomorphism. By Nakayama's lemma, 1.3, ϕ is surjective. Consider the exact sequence

$$0 \longrightarrow K \xrightarrow{\ \psi\ } L \xrightarrow{\ \phi\ } M \longrightarrow 0$$

to see that $\operatorname{Tor}_2^A(k, M) \simeq \operatorname{Tor}_1^A(k, K) = 0$. From this it follows that K is a free nonzero module. Note that $1 \otimes \psi : k \otimes_A K \to k \otimes_A L$ is the zero map. From this follows that

$$\operatorname{Ext}^i(1, \psi) : \operatorname{Ext}_A^i(k, K) \to \operatorname{Ext}_A^i(k, L)$$

is the zero map. We now get short exact sequences

$$0 \longrightarrow \operatorname{Ext}_A^i(k, L) \longrightarrow \operatorname{Ext}_A^i(k, M) \longrightarrow \operatorname{Ext}_A^{i+1}(k, K) \longrightarrow 0,$$

from which follow that $\operatorname{depth}(M) = \operatorname{depth}(A) - 1$. Suppose now that $\operatorname{proj dim}(M) = d \geq 2$. Choose an exact sequence

$$0 \longrightarrow N \longrightarrow L \longrightarrow M \longrightarrow 0,$$

where L is a finitely generated free module. From the long exact sequence in $\operatorname{Tor}_\cdot^A(k, -)$ follows that

$$\operatorname{proj dim}(N) = d - 1$$

and whence by the induction hypothesis

$$\operatorname{depth}(N) = \operatorname{depth}(A) - (d - 1).$$

In particular $\operatorname{depth}(N) < \operatorname{depth}(A)$. Using this inequality and the long exact $\operatorname{Ext}_A^\cdot(k, -)$ sequence arising from

$$0 \longrightarrow N \longrightarrow L \longrightarrow M \longrightarrow 0$$

it is easily seen that $\operatorname{depth}(N) = \operatorname{depth}(M) + 1$. $\qquad\square$

Corollary 2.48. *Suppose A is a Cohen–Macaulay local ring and let $M \neq 0$ be a module which has a resolution*

$$0 \longrightarrow L_r \longrightarrow L_{r-1} \longrightarrow \cdots \longrightarrow L_0 \longrightarrow M \longrightarrow 0$$

by finitely generated free modules L_i. Then

$$\dim(A) \leq r + \dim(M).$$

If the equality holds, then M is a Cohen–Macaulay module with projective dimension $\operatorname{proj dim}(M) = r$.

Proof. We have inequalities

$$\operatorname{proj dim}(M) \leq r \quad \text{and} \quad \operatorname{depth}(M) \leq \dim(M)$$

and the equalities

$$\operatorname{proj dim}(M) + \operatorname{depth}(M) = \operatorname{depth}(A) = \dim(A). \qquad \square$$

Corollary 2.49. *Suppose A is a regular local ring and $M \neq 0$ a finitely generated module. Then M is a Cohen–Macaulay module if and only if*

$$\operatorname{proj dim}(M) = \dim(A) - \dim(M).$$

In particular a Cohen–Macaulay module of dimension $\dim(A)$ is a free module.

Proof. Use 2.42 and 2.48. $\qquad \square$

2.10   ℑ-depth

Throughout this section A denotes a noetherian ring.

Lemma 2.50. *Let A be a noetherian local ring, $\mathfrak{I}$ an ideal contained in the maximal ideal of A and $M \neq 0$ a finitely generated A-module. Then*

$$\operatorname{Ext}_A^i(A/\mathfrak{I}, M) \neq 0, \qquad \text{for some } i \in \mathbb{N}.$$

Proof. Let us first note that

$$\operatorname{Hom}_A(A/\mathfrak{I}, M) \neq 0$$

if and only if no element of $\mathfrak{I}$ is a nonzero divisor for M, as it follows from 2.15

$$\operatorname{Ass}(\operatorname{Hom}_A(A/\mathfrak{I}, M)) = V(\mathfrak{I}) \cap \operatorname{Ass}(M)$$

and

$$V(\mathfrak{I}) \cap \operatorname{Ass}(M) \neq \emptyset \quad \text{if and only if} \quad \mathfrak{I} \subseteq \bigcup_{\mathfrak{p} \in \operatorname{Ass}(M)} \mathfrak{p}.$$

The proof can now be concluded as that of 2.21. $\qquad \square$

Definition 2.51. Let A be a noetherian local ring, $\mathfrak{I}$ an ideal contained in the maximal ideal of A and $M \neq 0$ a finitely generated A-module. Then we define the $\mathfrak{I}$-*depth* of M

$$\mathrm{depth}_{\mathfrak{I}}(M) = \inf\{i \in \mathbb{N} \mid \mathrm{Ext}_A^i(A/\mathfrak{I}, M) \neq 0\}.$$

Proposition 2.52. *Let A be a noetherian local ring, $\mathfrak{I}$ an ideal contained in the maximal ideal of A and $M \neq 0$ a finitely generated A-module. Then*

(1) $\mathrm{depth}_{\mathfrak{I}}(M) = 0$ *if and only if no element of $\mathfrak{I}$ is a nonzero divisor for M*

(2) *If $a \in \mathfrak{I}$ is a nonzero divisor for M, then*

$$\mathrm{depth}_{\mathfrak{I}}(M/aM) = \mathrm{depth}_{\mathfrak{I}}(M) - 1.$$

(3) $\mathrm{depth}_{\mathfrak{I}}(M)$ *is the maximal length of any M-regular sequence contained in $\mathfrak{I}$.*

Proof. Follows from the proof of 2.50. □

Proposition 2.53. *Let A be a noetherian local ring, $\mathfrak{I}$ an ideal contained in the maximal ideal of A and $M \neq 0$ a finitely generated A-module. Then*

$$\mathrm{depth}_{\mathfrak{I}}(M) = \inf_{\mathfrak{p} \in V(\mathfrak{I}) \cap \mathrm{Supp}(M)} \mathrm{depth}_{A_{\mathfrak{p}}}(M_{\mathfrak{p}}).$$

Proof. Let us first see that the two terms are different from zero at the same time. We have

$$
\begin{aligned}
\mathrm{depth}_{\mathfrak{I}}(M) \neq 0 \quad &\Leftrightarrow \quad \mathrm{Ass}(\mathrm{Hom}_A(A/\mathfrak{I}, M) = \emptyset \\
&\Leftrightarrow \quad V(\mathfrak{I}) \cap \mathrm{Ass}(M) = \emptyset \\
&\Leftrightarrow \quad \mathrm{depth}_{A_{\mathfrak{p}}}(M_{\mathfrak{p}}) \neq 0, \quad \mathfrak{p} \in V(\mathfrak{I}) \cap \mathrm{Supp}(M).
\end{aligned}
$$

Conclusion by induction on $\mathrm{depth}_{\mathfrak{I}}(M)$. □

Corollary 2.54. *Let $M \neq 0$ be a finitely generated A-module, $\mathfrak{I}$ and $\mathfrak{J}$ two ideals contained in the maximal ideal with $V(\mathfrak{I}) = V(\mathfrak{J})$. Then*

$$\mathrm{depth}_{\mathfrak{I}}(M) = \mathrm{depth}_{\mathfrak{J}}(M).$$

Proof. Follows from 2.53. □

Corollary 2.55. *Suppose that A is a Cohen–Macaulay local ring, then*

$$\operatorname{depth}_{\mathfrak{J}}(A) = \dim(A) - \dim(A/\mathfrak{J}).$$

Proof. For any prime ideal $\mathfrak{p}$ in A we have by 2.30 that

$$\dim(A) = \dim(A_{\mathfrak{p}}) + \dim(A/\mathfrak{p}).$$

Thus

$$
\begin{aligned}
\operatorname{depth}_{\mathfrak{J}}(A) &= \inf_{\mathfrak{p} \in V(\mathfrak{J})} \operatorname{depth}(A_{\mathfrak{p}}) \\
&= \inf_{\mathfrak{p} \in V(\mathfrak{J})} \dim(A_{\mathfrak{p}}) \\
&= \dim(A) - \sup_{\mathfrak{p} \in V(\mathfrak{J})} \dim(A/\mathfrak{p}) \\
&= \dim(A) - \dim(A/\mathfrak{J}). \qquad \square
\end{aligned}
$$

Corollary 2.56. *Suppose A is a regular local ring and M a finitely generated module with $\mathfrak{J} = \operatorname{Ann}(M) \neq A$. Then M is a Cohen–Macaulay module if and only if*

$$\operatorname{proj} \dim(M) = \operatorname{depth}_{\mathfrak{J}}(A).$$

Proof. Use 2.55 and 2.49. $\qquad \square$

2.11   The acyclicity theorem

Throughout this section A denotes a noetherian local ring, $X = \operatorname{Spec}(A)$. For a nonempty closed set $Z = V(\mathfrak{J})$ of X define $\dim(Z) = \dim(A/\mathfrak{J})$ and $\operatorname{codim}(Z) = \dim(X) - \dim(Z)$.

Definition 2.57. Let $L.$ be a bounded complex of finitely generated A-modules. Define the *support* of $L.$ by

$$\operatorname{Supp}(L.) = \{\mathfrak{p} \in \operatorname{Spec}(A) \mid H.(L. \otimes_A A_{\mathfrak{p}}) \neq 0\}.$$

Lemma 2.58. $\operatorname{Supp}(L.)$ *is a closed set.*

Proof. It follows easily from Nakayama's lemma, 1.3, that $\operatorname{Supp}(L.) = \bigcup_i \operatorname{Supp}(H_i(L.))$. $\qquad \square$

Theorem 2.59 (acyclicity theorem). *Let*

$$0 \longrightarrow L_s \stackrel{\partial_s}{\longrightarrow} L_{s-1} \longrightarrow \cdots \longrightarrow L_0 \longrightarrow 0$$

be a complex of finitely generated A-modules. Assume for $i = 1, \ldots, s$

(1) $L_i \neq 0$ *and* $\mathrm{depth}(L_i) \geq i$.

(2) $H_i(L.) = 0$ *or* $\mathrm{depth}(H_i(L.)) = 0$.

Then

$$H_i(L.) = 0 \qquad \text{for } i \geq 1.$$

Proof. We are going to prove that

$$H_i(L.) = 0 \quad \text{and} \quad \mathrm{depth}(\mathrm{Cok}(\partial_{i+1})) \geq i, \qquad i = 1, \ldots, s.$$

This will be done by decreasing induction. The case $i = s$ is clear. Note that $H_i(L.)$ is a submodule of $\mathrm{Cok}(\partial_{i+1})$, thus if $H_i(L.) \neq 0$ we would have $\mathrm{depth}(\mathrm{Cok}(\partial_{i+1})) = 0$. Next consider the exact sequence

$$0 \longrightarrow \mathrm{Cok}(\partial_{i+1}) \longrightarrow L_{i-1} \longrightarrow \mathrm{Cok}(\partial_i) \longrightarrow 0.$$

The long exact $\mathrm{Ext}_A^{\bullet}(k, -)$ sequence and the induction hypothesis implies $\mathrm{depth}(\mathrm{Cok}(\partial_i)) \geq i - 1$. $\qquad\square$

Corollary 2.60. *Let* $\mathrm{depth}(A) = s$ *and let*

$$0 \longrightarrow L_s \stackrel{\partial_s}{\longrightarrow} L_{s-1} \longrightarrow \cdots \longrightarrow L_0 \longrightarrow 0$$

be a complex of finitely generated free A-modules. If

$$H_i(L.) = 0 \quad \text{or} \quad \mathrm{depth}(H_i(L.)) = 0 \qquad \text{for } i \geq 1$$

then

$$H_i(L.) = 0 \qquad \text{for } i \geq 1.$$

Proof. Straightforward from 2.59. $\qquad\square$

Corollary 2.61. *Let A be a Cohen–Macaulay local ring with $\dim(A) = s$ and let*

$$0 \longrightarrow L_s \overset{\partial_s}{\longrightarrow} L_{s-1} \longrightarrow \cdots \longrightarrow L_0 \longrightarrow 0$$

be a complex of finitely generated free A-modules. If

$$H_i(L.) \text{ has finite length for } i \geq 1$$

then

$$H_i(L.) = 0 \qquad \text{for } i \geq 1.$$

Proof. Straightforward from 2.59. $\qquad\qquad\qquad\qquad\qquad\qquad\square$

Corollary 2.62. *Let A be a Cohen–Macaulay local ring and*

$$0 \longrightarrow L_s \overset{\partial_s}{\longrightarrow} L_{s-1} \longrightarrow \cdots \longrightarrow L_0 \longrightarrow 0$$

a complex of finitely generated free A-modules. Suppose $0 \neq H.(L.)$ has finite length. Then $s \geq \dim(A)$.

Proof. Straightforward from 2.59. $\qquad\qquad\qquad\qquad\qquad\qquad\square$

Proposition 2.63. *Let A be a Cohen–Macaulay local ring and*

$$0 \longrightarrow L_s \overset{\partial_s}{\longrightarrow} L_{s-1} \longrightarrow \cdots \longrightarrow L_0 \longrightarrow 0$$

a complex of finitely generated free A-modules. If $\mathrm{Supp}(L.) \neq \emptyset$, then

$$\mathrm{codim}(\mathrm{Supp}(L.)) \leq s.$$

If $\mathrm{codim}(\mathrm{Supp}(L.)) = s$ then $H_i(L.) = 0$ for $i \geq 1$ and $H_0(L.)$ is a Cohen–Macaulay module of projective dimension s. Moreover $L.$ above is a projective resolution of $H_0(L.)$.

Proof. Let $\mathfrak{I}$ be an ideal such that $V(\mathfrak{I}) = \mathrm{Supp}(L.)$. Then we have $\mathrm{depth}_{\mathfrak{I}}(H_i(L.)) = 0$ for $i = 0, \ldots, s$. Moreover we have by 2.55

$$\mathrm{depth}_{\mathfrak{I}}(A) = \mathrm{codim}(\mathrm{Supp}(L.) \quad (= r).$$

Proceeding precisely as in the proof of 2.59 (replacing depth by $\mathrm{depth}_{\mathfrak{I}}$) one sees that $H_i(L.) = 0$ for $i > s - r$. This implies $r \leq s$. In case $r = s$ we have $\dim(H_0(L.)) + s = \dim(A)$, since $\mathrm{Supp}(H_0(L.)) = \mathrm{Supp}(L.)$, conclusion by 2.48. $\qquad\qquad\square$

Corollary 2.64. *Let A be a Cohen–Macaulay local ring and $\mathfrak{I}$ a proper ideal. Let*

$$0 \longrightarrow L_s \xrightarrow{\partial_s} L_{s-1} \longrightarrow \cdots \longrightarrow L_0 \longrightarrow 0$$

be a complex of finitely generated free A-modules such that $H_0(L.) \simeq A/\mathfrak{I}$. If $\operatorname{codim}(\operatorname{Supp}(L.)) = s$ then $H_i(L.) = 0$ for $i \geq 1$ and $A/\mathfrak{I}$ is a Cohen–Macaulay module of projective dimension s. Moreover $L.$ above is a projective resolution of $A/\mathfrak{I}$. The module

$$\operatorname{Ext}_A^s(A/\mathfrak{I}, A)$$

is a Cohen–Macaulay module of finite projective dimension s and

$$\operatorname{Ext}_A^s(\operatorname{Ext}_A^s(A/\mathfrak{I}, A), A) \simeq A/\mathfrak{I}.$$

Proof. From 2.63 follows that the dual complex

$$0 \longrightarrow \operatorname{Hom}_A(L_0, A) \longrightarrow \cdots \longrightarrow \operatorname{Hom}_A(L_s, A) \longrightarrow 0$$

becomes a free resolution of $\operatorname{Ext}_A^s(A/\mathfrak{I}, A)$, giving the claims. $\qquad\square$

2.12 An example

Example 2.65. Let A denote a ring, E a free A-module of rank n and $\phi : E \to E$ an A-linear endomorphism. The cofactor of ϕ, $\tilde{\phi} : E \to E$, is defined as usual from a matrix representation of ϕ. $\tilde{\phi}$ is independent of chosen basis and satisfies

$$\phi \circ \tilde{\phi} = \tilde{\phi} \circ \phi = \det(\phi)\, \mathbf{1}_E .$$

Consider the complex

$$A \xrightarrow{a \mapsto (a\,\mathbf{1}_E, a\,\mathbf{1}_E)} \operatorname{End}(E) \oplus \operatorname{End}(E) \xrightarrow{(\phi,\psi) \mapsto \operatorname{Tr}(\phi-\psi)} A$$

and let $\operatorname{End}(E)\tilde{\oplus}\operatorname{End}(E)$ denote the middle homology module of this complex. Let

$$C. = \quad C_4 \longrightarrow C_3 \longrightarrow C_2 \longrightarrow C_1 \longrightarrow C_0 \qquad (2.1)$$

denote the following complex

$$A \xrightarrow{a \mapsto a\tilde{\phi}} \operatorname{End}(E) \xrightarrow{\psi \mapsto (\phi\psi, \psi\phi)} \operatorname{End}(E) \tilde{\oplus} \operatorname{End}(E)$$
$$\xrightarrow{(\alpha,\beta) \mapsto (\alpha\phi - \phi\beta)} \operatorname{End}(E) \xrightarrow{\psi \mapsto \operatorname{Tr}(\psi\tilde{\phi})} A.$$

Let $\mathfrak{I}_{n-1}$ denote the ideal generated by all $(n-1)$-minors of ϕ. We have

$$H_0(C.) = A/\mathfrak{I}_{n-1}$$

and we are going to prove that

$$\operatorname{Supp}(C.) = V(\mathfrak{I}_{n-1}). \tag{2.2}$$

For this let us note that

$$\Sigma_i (-1)^i \operatorname{rank}(C_i) = 0$$

and that the isomorphism

$$\operatorname{End}(E) \to \operatorname{End}(E)^{\vee}, \qquad \alpha \mapsto (\beta \mapsto \operatorname{Tr}(\beta\alpha))$$

induces an isomorphism

$$C. \simeq C.^{\vee}[-4].$$

To establish (2.2) it suffices to prove that if $A = k$ is a field and $\tilde{\phi} \neq 0$ then $H.(C.) = 0$. From the above it clearly remains to prove that $H_3(C.) = 0$. Given a cycle $\psi \in C_3$, i.e., there exists $x \in k$ such that $\phi\psi = \psi\phi = x\, 1_E$. We must show that ψ is a scalar multiple of $\tilde{\phi}$. In case $\det(\phi) \neq 0$ this follows from the cofactor equation. Suppose $\det(\phi) = 0$, then we get from $\det(\psi) \det(\phi) = x^n$ that $x = 0$. Thus it will suffice to prove that

$$\operatorname{rank}(\{ \omega \in \operatorname{End}(E) \mid \phi\omega = \omega\phi = 0\}) = 1.$$

So suppose $\phi\omega = \omega\phi = 0$, which means that ω can be factored

$$E \longrightarrow \operatorname{Cok}(\phi) \longrightarrow \operatorname{Ker}(\phi) \longrightarrow E$$

but $\tilde{\phi} \neq 0$ whence $\operatorname{rank} \operatorname{Cok} \phi = \operatorname{rank} \operatorname{Ker} \phi = 1$.

     The preceding construction is due to Gulliksen–Negaard. As an application of 2.63 we get the following proposition.

Proposition 2.66. *Let A be a Cohen–Macaulay local ring with residue field k and $\phi : E \to E$ an endomorphism of a finitely generated free A-module of rank n such that $\phi \otimes 1 : E \otimes_A k \to E \otimes_A k$ has rank $< n - 1$. Then we have for the ideal $\mathfrak{I}_{n-1}$ generated by the $(n-1)$-minors of ϕ*

$$\mathrm{codim}(V(\mathfrak{I}_{n-1})) \leq 4.$$

In case of equality, $A/\mathfrak{I}_{n-1}$ is a Cohen–Macaulay module of projective dimension 4 with $C.$, (2.1), being a free resolution.

Chapter 3

Divisor Theory

3.1 Discrete valuation rings

Definition 3.1. A *discrete valuation ring* is a noetherian local domain in which the maximal ideal is principal and nonzero. That is the same as a regular local ring of dimension 1, 1.9.

Proposition 3.2. *Let A be a discrete valuation ring with maximal ideal $\mathfrak{m}$. The only nonzero ideals in A are those of the form $\mathfrak{m}^i$, $i \in \mathbb{N}$.*

Proof. Let $a \in A - 0$. Since, by Krull's intersection theorem, 1.35, $\bigcap_i \mathfrak{m}^i = 0$, we can choose $i \in \mathbb{N}$ such that $a \in \mathfrak{m}^i - \mathfrak{m}^{i+1}$. Let π generate $\mathfrak{m}$, and write $a = u\pi^i$, where u is a unit. Conclude by noting that any ideal is a sum of principal ideals. $\qquad\square$

Definition 3.3. Let A denote a discrete valuation ring with maximal ideal $\mathfrak{m}$ and fraction field K. For an element f in $K^* = K - \{0\}$ write $f = u\pi^i$ where π is a generator for $\mathfrak{m}$, u a unit in A and $i \in \mathbb{Z}$. The integer i is easily seen to be independent of the choice of π, and we write $\nu_A(f) = i$. The map

$$\nu_A : K^* \to \mathbb{Z}$$

is a morphism from the multiplicative group K^* to $\mathbb{Z}$ and it is called the *valuation of K associated A*. Note

Remark 3.4.

$$\nu_A(a) = \ell_A(A/(a)), \qquad a \in A - \{0\}. \tag{3.1}$$

Proposition 3.5. *Let E denote a finitely generated free module over the discrete valuation ring A and $f : E \to E$ an injective A-linear endomorphism. Then*

$$\nu_A(\det(f)) = \ell_A(\mathrm{Cok}(f)).$$

Proof. If g is a second injective endomorphism on E then

$$\nu_A(\det(g \circ f)) = \nu_A(\det(g)) + \nu_A(\det(f))$$
$$\ell_A(\mathrm{Cok}(g \circ f)) = \ell_A(\mathrm{Cok}(g)) + \ell_A(\mathrm{Cok}(f)).$$

By a row and column operations on matrices over A we can find automorphisms g, h of E such that $g \circ f \circ h$ is diagonal. The proposition now follows from formula (3.1).   □

3.2   Normal domains

Throughout this section A will denote an integral domain with fraction field K.

Definition 3.6. The integral domain A is said to be a *normal domain* if A is integrally closed in K. i.e., any element f in K that satisfies an equation of the form

$$f^n + a_1 f^{n-1} + \cdots + a_n = 0, \qquad a_i \in A$$

belongs to A.

Remark 3.7. If A is a normal domain and S a multiplicative subset of A, not containing 0, then $S^{-1}A$ is a normal domain.

Lemma 3.8. *Let A denote a normal domain, $\mathfrak{q} \neq 0$ a finitely generated ideal. For $a, b \in A$*

$$a\mathfrak{q} \subseteq b\mathfrak{q} \quad \Rightarrow \quad (a) \subseteq (b).$$

Proof. We may assume $b \neq 0$ and put $f = a/b$. We have $\mathfrak{q}f \subseteq \mathfrak{q}$ and want to show that $f \in A$. Let $x_1, \ldots, x_d$ generate $\mathfrak{q}$. We can find a $d \times d$-matrix U with entries in A such that

$$f \begin{pmatrix} x_1 \\ \vdots \\ x_d \end{pmatrix} = U \begin{pmatrix} x_1 \\ \vdots \\ x_d \end{pmatrix}.$$

This implies that

$$\det(U - fI) = 0.$$

This is an integral equation for f and whence $f \in A$.   □

Theorem 3.9 (Serre's criterion). *Let A be a noetherian domain. Then A is a normal domain if and only if the following two conditions are satisfied for prime ideals $\mathfrak{p}$ in A*

(1) *If $\dim(A_\mathfrak{p}) = 1$, then $A_\mathfrak{p}$ is a discrete valuation ring.*

(2) *If $\mathfrak{p} \in \mathrm{Ass}(A/(b))$ for some $b \in A - \{0\}$, then $\dim(A_\mathfrak{p}) = 1$.*

Proof. Suppose A is a normal domain, and let the prime ideal $\mathfrak{p}$ be associated to $A/(b)$, $b \neq 0$. We are going to prove that $A_\mathfrak{p}$ is a discrete valuation ring. By 3.7 and 2.13 we may assume that A is local with maximal ideal $\mathfrak{p}$. Choose $a \in A$ such that

$$a\mathfrak{p} \subseteq (b) \quad \text{and} \quad (a) \not\subseteq (b).$$

The second condition implies by 3.8 that

$$a\mathfrak{p} \not\subseteq b\mathfrak{p}.$$

Choose $x \in a\mathfrak{p}$ with $x \notin b\mathfrak{p}$ and write

$$x = \pi a = ub, \qquad \pi \in \mathfrak{p},\ u \text{ unit in } A.$$

This gives rise to inclusions

$$a\mathfrak{p} \subseteq (b) \subseteq a(\pi).$$

It follows $\mathfrak{p} = (\pi)$.

Conversely, suppose conditions (1) and (2) are satisfied. As a discrete valuation ring is normal and the intersection of normal domains is again normal it suffices to prove the following observation.

Observation 3.10.

$$A = \bigcap_\mathfrak{p} A_\mathfrak{p}, \qquad \mathfrak{p} \text{ prime ideal such that } \dim(A_\mathfrak{p}) = 1. \tag{3.2}$$

Given $0 \neq f \in \bigcap_\mathfrak{p} A_\mathfrak{p}$. We may write $f = a/b$ and consider the A-module $N = (a,b)/(b)$. We want to show that $N = 0$, or what amounts to the same by 2.7, that $\mathrm{Ass}(N) = \emptyset$. Note by assumption $N_\mathfrak{p} = 0$ for all $\mathfrak{p}$ with $\dim(A_\mathfrak{p}) = 1$. On the other hand, since $N \subseteq A/(b)$ it follows that $\mathfrak{p} \in \mathrm{Ass}(N)$ implies $\dim(A_\mathfrak{p}) = 1$. □

Remark 3.11. By 3.10

$$A = \bigcap_\mathfrak{p} A_\mathfrak{p}, \qquad \mathfrak{p} \text{ prime ideal such that } \dim(A_\mathfrak{p}) = 1$$

for any noetherian normal domain A.

Remark 3.12. By means of the concept of depth we can restate Serre's criterion for normality.

Let A be a noetherian domain. Then A is a normal domain if and only if the following conditions are satisfied

(1) $\dim(A_{\mathfrak{p}}) = 1 \Rightarrow A_{\mathfrak{p}}$ is a discrete valuation ring.

(2) $\dim(A_{\mathfrak{p}}) \geq 2 \Rightarrow \operatorname{depth}(A_{\mathfrak{p}}) \geq 2$.

3.3   Divisors

Let A denote a noetherian normal domain with fraction field K. A prime ideal $\mathfrak{p}$ of A with $\dim(A_{\mathfrak{p}}) = 1$ is called a *prime divisor* and the corresponding valuation for $A_{\mathfrak{p}}$ of K will be denoted by

$$\nu_{\mathfrak{p}} : K^* \to \mathbb{Z}.$$

The free abelian group on the prime divisors will be called the *divisor group* of A and will be denoted $\operatorname{Div}(A)$.

Proposition 3.13. *Let $f \in K^*$, then $\nu_{\mathfrak{p}}(f) = 0$ except for finitely many prime divisors.*

Proof. Writing $f = a/b$ with $a, b \in A$, it suffices to treat the case $f \in A - \{0\}$. In that case $\nu_{\mathfrak{p}}(f) \neq 0$ implies $\mathfrak{p} \in \operatorname{Ass}(A/(f))$ as it follows from formula (3.1). Conclusion by 2.11. $\qquad\square$

Definition 3.14. For $f \in K^*$ define the *divisor* of f, $\operatorname{div}(f) \in \operatorname{Div}(A)$, by

$$\operatorname{div}(f) = \sum_{\mathfrak{p}} \nu_{\mathfrak{p}}(f)\,\mathfrak{p}, \qquad \mathfrak{p} \text{ being a prime divisor.}$$

In this way we get the *divisor map*

$$\operatorname{div} : K^* \to \operatorname{Div}(A).$$

Definition 3.15. A divisor $D = \sum n_{\mathfrak{p}}\mathfrak{p}$ is called *positive*, $D \geq 0$ if $n_{\mathfrak{p}} \geq 0$ for all prime divisors $\mathfrak{p}$.

Theorem 3.16. *Let $f \in K^*$. Then the divisor of f is positive, $\operatorname{div}(f) \geq 0$, if and only if $f \in A$.*

Proof. Note that if $\mathfrak{p}$ is a prime divisor, then

$$A_{\mathfrak{p}} - \{0\} = \{f \in K^* \mid \nu_{\mathfrak{p}}(f) \geq 0\}.$$

Conclusion by formula (3.2). $\qquad\square$

Definition 3.17. The cokernel of the divisor map

$$\mathrm{div} : K^* \to \mathrm{Div}(A)$$

is called the *class group* of A and will be denoted $\mathrm{Cl}(A)$. The canonical projection will be denoted $c : \mathrm{Div}(A) \to \mathrm{Cl}(A)$. Let $U(A)$ denote the group of multiplicatively invertible elements in A. It follows from 3.16 that we have a canonical exact sequence

$$1 \longrightarrow U(A) \longrightarrow K^* \longrightarrow \mathrm{Div}(A) \longrightarrow \mathrm{Cl}(A) \longrightarrow 0.$$

Let us note the following consequence of 3.16.

Corollary 3.18. *Let $\mathfrak{p}$ be a prime divisor. Then $\mathfrak{p}$ is a principal ideal if and only if $c(\mathfrak{p}) = 0$.*

Proof. Suppose $c(\mathfrak{p}) = 0$ and choose $f \in K^*$ with $\mathrm{div}(f) = \mathfrak{p}$. By 3.16 we have $f \in \mathfrak{p}$. Suppose $0 \neq g \in \mathfrak{p}$. Then $\mathrm{div}(g/f) \geq 0$ and whence $g \in (f)$. $\square$

3.4 Unique factorization

Let us recall the classical theory for factorization in a noetherian domain A. Let $\mathcal{P}$ denote the set of principal ideals distinct form (0) and (1) ordered by inclusion. An element $p \in \mathcal{P}$ is called *irreducible* if it is maximal in that set. It follows easily from the noetherian hypothesis that any $p \in \mathcal{P}$ can be factored in irreducible elements. Let us call A a *unique factorization domain* if the factorization into irreducible elements is unique up to order of the factors. It remains the same to require that if $p \in \mathcal{P}$ is irreducible, the p is a prime ideal.

Proposition 3.19. *Let A be a noetherian domain. Then A is a unique factorization domain if and only if A is a normal domain with trivial class group, $\mathrm{Cl}(A) = 0$.*

Proof. Suppose A is a unique factorization domain. Then an elementary consideration shows that A is a normal domain. Let $\mathfrak{p}$ be a prime ideal in A with $\dim(A_\mathfrak{p}) = 1$, then $\mathfrak{p}$ is principal: choose any $0 \neq d \in \mathfrak{p}$, factor d in irreducible elements to see that $\mathfrak{p}$ contains an irreducible element π, which generates a prime ideal (π), whence $(\pi) = \mathfrak{p}$ and therefore $\operatorname{div}(\pi) = \mathfrak{p}$ and so $\operatorname{Cl}(A) = 0$.

Assume A is normal with $\operatorname{Cl}(A) = 0$. For each prime divisor $\mathfrak{p}$ choose $f_\mathfrak{p} \in K$ with $\operatorname{div}(f_\mathfrak{p}) = \mathfrak{p}$. It follows from 3.16 that $f_\mathfrak{p} \in A$ and that $(f_\mathfrak{p})$ is irreducible. Let now $a \in A - 0$ and consider the element

$$f = \prod_\mathfrak{p} f_\mathfrak{p}^{\nu_\mathfrak{p}(a)},$$

which has the same divisor as a. It follows that each principal ideal in A can be decomposed in a unique way into a product of the irreducible ideals $(f_\mathfrak{p})$. □

3.5   Torsion modules

Let A be a domain with fraction field K. An A-module M is called a *torsion module* if each element of M has a nonzero annihilator. M is a torsion module if and only if $M \otimes_A K = 0$.

Throughout this section we shall let A denote a noetherian normal domain with fraction field K.

Definition 3.20. Let M be a finitely generated torsion A-module. Each prime divisor $\mathfrak{p}$ in A is either not in $\operatorname{Supp}(M)$ or minimal in $\operatorname{Supp}(M)$. So the $A_\mathfrak{p}$-module $M_\mathfrak{p}$ is of finite length and moreover $\ell_{A_\mathfrak{p}}(M_\mathfrak{p}) = 0$ except for finitely many $\mathfrak{p}$'s, 2.11 and 2.14. We define the *divisor* $h(M)$ by

$$h(M) = \sum_\mathfrak{p} \ell_{A_\mathfrak{p}}(M_\mathfrak{p})\,\mathfrak{p}.$$

Remark 3.21. Given a short exact sequence $0 \to N \to M \to P \to 0$. M is a finitely generated torsion module if and only if N and P are finitely generated torsion modules. In the torsion case

$$h(M) = h(N) + h(P).$$

Remark 3.22. For a prime divisor $\mathfrak{p}$ of A we have $h(A/\mathfrak{p}) = \mathfrak{p}$.

Proposition 3.23. *Let L be a finitely generated free A-module. For an injective endomorphism $\phi : L \to L$, we have*

$$h(\mathrm{Cok}(\phi)) = \mathrm{div}(\det(\phi)).$$

Proof. Follows immediately from 3.5. $\qquad\qquad\square$

3.6 The first Chern class

Throughout this section A denotes a noetherian normal domain with fraction field K. The map defined below is called the *first Chern class*.

Theorem 3.24. *There exists one and only one map*

$$c_1 : (\textit{finitely generated } A\textit{-modules}) \to \mathrm{Cl}(A)$$

such that

(1) *Given a short exact sequence $0 \to N \to M \to P \to 0$ of finitely generated modules, then*

$$c_1(M) = c_1(N) + c_1(P).$$

(2) *For any free module L, $c_1(L) = 0$.*

(3) *For any torsion module M, $c_1(M) = c(h(M))$.*

Proof. Uniqueness of c_1 follows immediately from the first part of 3.25 below.

Let us prove that if M is a finitely generated module and L and L' are free submodules with M/L and M/L' torsion modules, then

$$c(h(M/L)) = c(h(M/L')).$$

By the second part of 3.25, we may assume $L' \subseteq L$. We have

$$h(M/L') = h(M/L) + h(L/L').$$

Choose an isomorphism $L \simeq L'$ and let $\phi : L \to L$ denote the composition with the inclusion. Then

$$h(L/L') = h(\mathrm{Cok}(\phi)) = \mathrm{div}(\det(\phi)),$$

which proves the assertion. We may now define

$$c_1(M) = c(h(M/L)), \qquad L \subseteq M, \ L, \ M/L \text{ torsion.}$$

(1) 'additivity' of c_1 follows from the third part of 3.25. (2) and (3) are clearly satisfied. □

Lemma 3.25. *Let A denote an integral domain.*

(1) *For any finitely generated A-module M there exists a free submodule $L \subseteq M$ such that M/L is a torsion module.*

(2) *If L' and L'' are submodules as in (1) then there exists a third L with the same properties and contained in both L' and L''.*

(3) *Given an exact sequence $0 \to M' \to M \to M'' \to 0$ of finitely generated A-modules, then there exists a commutative diagram with exact rows*

$$
\begin{array}{ccccccccc}
0 & \longrightarrow & L' & \longrightarrow & L & \longrightarrow & L'' & \longrightarrow & 0 \\
& & \downarrow & & \downarrow & & \downarrow & & \\
0 & \longrightarrow & M' & \longrightarrow & M & \longrightarrow & M'' & \longrightarrow & 0,
\end{array}
$$

where the first row consists of finitely generated free modules and the vertical maps are injections with torsion cokernels.

Proof. (1) Elements $x_1, \ldots, x_n$ giving a K-basis in $M \otimes_A K$ generate a free A-submodule with torsion cokernel.

(2) Use (1) on the A-module $L' \cap L''$.

(3) Supply a K-basis for $M' \otimes_A K$ to a basis for $M \otimes_A K$. □

Proposition 3.26. *A normal noetherian domain with the property that any finitely generated module admits a finite resolution by finitely generated free modules is a unique factorization domain.*

Proof. Note that c_1 is surjective as it follows from 3.24 and 3.22. Let M be a module with a resolution

$$0 \longrightarrow L_n \longrightarrow L_{n-1} \longrightarrow \cdots \longrightarrow L_0 \longrightarrow M \longrightarrow 0,$$

where the L_i's are finitely generated free A-modules. Then by induction on n

$$c_1(M) = \sum_i (-1)^i c_1(L_i) = 0. \qquad \Box$$

3.7 Regular local rings

Theorem 3.27 (Auslander–Buchsbaum). *A regular local ring is a unique factorization domain.*

Proof. Suppose $d = \dim(A) \geq 2$. Since $\mathfrak{m}$ is not free

$$\mathrm{Tor}_2^A(k, k) \simeq \mathrm{Tor}_1^A(k, \mathfrak{m}) \neq 0.$$

Let $0 \neq a \in \mathfrak{m}$. If $\mathfrak{m} \in \mathrm{Ass}(A/(a))$. Then there exists an exact sequence

$$0 \longrightarrow k \longrightarrow A/(a) \longrightarrow E \longrightarrow 0$$

giving us by 2.45

$$\mathrm{Tor}_{i+1}^A(k, E) \simeq \mathrm{Tor}_i^A(k, k), \qquad i \geq 2.$$

Since there exists $r \geq 2$ such that $\mathrm{Tor}_{r+1}^A(k, k) = 0$, $\mathrm{Tor}_r^A(k, k) \neq 0$ it follows $\mathrm{Tor}_{r+1}^A(k, E) \neq 0$, a contradiction, so $\mathfrak{m} \notin \mathrm{Ass}(A/(a))$. It follows from Serre's criterion that A is a normal domain. Conclusion by 2.42 and 3.26. $\qquad\square$

3.8 Picard groups

Recall that for a ring A the *Picard group* of A, $\mathrm{Pic}(A)$ is the set of isomorphism classes of projective modules of constant rank 1 organized as an abelian group by the tensor product.

Proposition 3.28. *Let A be a noetherian normal domain. Then the restriction of the first Chern class*

$$c_1 : \mathrm{Pic}(A) \to \mathrm{Cl}(A)$$

is additive and injective. That is,

(1) $c_1(L \otimes_A L') = c_1(L) + c_1(L')$.
(2) $c_1(L) = 0 \;\Rightarrow\; L \simeq A$.

Proof. Choose an exact sequence

$$0 \longrightarrow A \longrightarrow L \longrightarrow C \longrightarrow 0$$

and apply $- \otimes_A L'$ to get an exact sequence

$$0 \longrightarrow L' \longrightarrow L \otimes_A L' \longrightarrow C \otimes_A L' \longrightarrow 0,$$

from which we get

$$c_1(L \otimes_A L') = c_1(L') + c_1(C \otimes_A L').$$

Note that $C \otimes_A L'$ is a torsion module, whence

$$c_1(C \otimes_A L') = c(h(C \otimes_A L')).$$

Note also

$$c_1(L) = c_1(C) = c(h(C)).$$

Thus it suffices to prove that $h(C) = h(C \otimes_A L')$. Let $\mathfrak{p}$ be a prime divisor, then

$$(C \otimes_A L')_{\mathfrak{p}} \simeq C_{\mathfrak{p}} \otimes_{\mathfrak{p}} L'_{\mathfrak{p}} \simeq C_{\mathfrak{p}} \otimes_{A_{\mathfrak{p}}} A_{\mathfrak{p}} \simeq C_{\mathfrak{p}}$$

and the equality $h(C) = h(C \otimes_A L')$ follows.

Let us next prove that $c_1 : \mathrm{Pic}(A) \to \mathrm{Cl}(A)$ is injective. So let L represent an element in $\mathrm{Pic}(A)$ with $c_1(L) = 0$. Choose an exact sequence

$$0 \longrightarrow A \overset{i}{\longrightarrow} L \overset{p}{\longrightarrow} C \longrightarrow 0,$$

note, $c_1(L) = c_1(C) = c(h(C))$, whence the divisor $h(C)$ is principal. Choose $a \in A - \{0\}$ such that $\mathrm{div}(a) = h(C)$ and denote by $a_L : L \to L$ multiplication by a. We are going to prove that

$$0 \longrightarrow L \overset{a_L}{\longrightarrow} L \overset{p}{\longrightarrow} C \longrightarrow 0$$

is exact, which will imply $L \simeq A$. The composite of p and a_L is zero. In fact

$$\mathrm{Ass}(\mathrm{Im}(p \circ a_L)) \subseteq \mathrm{Ass}(C)$$

thus it suffices to se that $p \circ a_L$ is zero after localization of a prime divisor, which is clear. Next let us prove that $H = \mathrm{Ker}(p)/\mathrm{Im}(a_L)$ is zero. Note $\mathrm{Ass}(H) \subseteq \mathrm{Ass}(\mathrm{Cok}(a_L))$ thus it suffices to check that after localization of H at a prime divisor is zero. $\qquad \square$

Proposition 3.29. *Let A denote a noetherian normal domain. Then the following conditions are equivalent,*

(1) *Any prime divisor $\mathfrak{p}$ is a projective module of rank one.*

(2) *$A_{\mathfrak{q}}$ is a unique factorization domain for all prime ideals $\mathfrak{q}$.*

(3) *$c_1 : \operatorname{Pic}(A) \to \operatorname{Cl}(A)$ is an isomorphism.*

Proof. (1) and (2) are equivalent in virtue of the fact that any finitely generated projective module over a local ring is free, 2.33. (1) implies (3) as it follows by considering the exact sequence

$$0 \longrightarrow \mathfrak{p} \longrightarrow A \longrightarrow A/\mathfrak{p} \longrightarrow 0,$$

which gives $c_1(\mathfrak{p}) = -c(\mathfrak{p})$. To prove that (3) implies (1) use the following lemma with $S = A - \mathfrak{q}$. □

Lemma 3.30. *Let S denote a multiplicative closed subset of the noetherian normal domain A. Then there is a canonical surjective map $r : \operatorname{Cl}(A) \to \operatorname{Cl}(S^{-1}A)$ making the following diagram commutative*

$$
\begin{array}{ccc}
\operatorname{Pic}(A) & \xrightarrow{\;c_1\;} & \operatorname{Cl}(A) \\
\downarrow & & \downarrow{\scriptstyle r} \\
\operatorname{Pic}(S^{-1}A) & \xrightarrow{\;c_1\;} & \operatorname{Cl}(S^{-1}A),
\end{array}
$$

where the vertical map to the left is $L \mapsto S^{-1}L$.

Proof. Let $\rho : \operatorname{Div}(A) \to \operatorname{Div}(S^{-1}A)$ denote the additive map given by

$$\rho(\mathfrak{p}) = \begin{cases} 0 & \text{if } \mathfrak{p} \cap S \neq \emptyset, \\ S^{-1}\mathfrak{p} & \text{if } \mathfrak{p} \cap S = \emptyset. \end{cases}$$

We have a commutative diagram

$$
\begin{array}{ccc}
K^* & \xrightarrow{\;\mathrm{div}\;} & \operatorname{Div}(A) \\
\downarrow{\scriptstyle \mathrm{id}} & & \downarrow{\scriptstyle \rho} \\
K^* & \xrightarrow{\;\mathrm{div}\;} & \operatorname{Div}(S^{-1}A).
\end{array}
$$

We let $r : \operatorname{Cl}(A) \to \operatorname{Cl}(S^{-1}A)$ denote the induced map. Clearly ρ is surjective and whence r is surjective. The commutativity is easily seen. □

3.9 Dedekind domains

A *Dedekind domain* is a noetherian normal domain in which all prime divisors are maximal ideals. A module over a domain is called *torsion free* if the annihilator of any nonzero element is zero.

Proposition 3.31. *Let A be a Dedekind domain.*

(1) *If A is a unique factorization domain then it is a principal ideal domain.*

(2) *If A has only finitely many prime divisors then it is a principal ideal domain.*

Proof. (1) For a nonzero ideal $\mathfrak{a}$ there is $a \in A$ such that $\operatorname{div}(a) = h(A/\mathfrak{a})$, 3.19. By localization $\mathfrak{a} = (a)$.

(2) Let $\mathfrak{p}_1, \ldots, \mathfrak{p}_s$ be the finitely many prime divisors. Choose $x_i \in \mathfrak{p}_i - \mathfrak{p}_1 \cup \cdots \cup \mathfrak{p}_i^2 \cup \cdots \cup \mathfrak{p}_s$. Then $\operatorname{div}(x_i) = \mathfrak{p}_i$. By 3.19 A is a unique factorization domain, so conclude by (1). □

Theorem 3.32. *Let A be a Dedekind domain and let M be a finitely generated A-module.*

(1) *If M is torsion free then there is a free submodule $L \subseteq M$ such that M/L is projective of rank one*

$$M \simeq L \oplus M/L.$$

(2) *The submodule $T \subseteq M$ of torsion elements has a torsion free complement $N \subseteq M$*
$$M = N \oplus T.$$

Proof. (1) Let A have fraction field K and assume $\operatorname{rank}_K M \otimes_A K = n$. By choosing a common denominator for a basis we may assume $M \subset A^n$. By induction on n, $M \simeq \mathfrak{a}_1 \oplus \cdots \oplus \mathfrak{a}_n$ for nonzero ideals $\mathfrak{a}_i$ in A. It suffices to treat the case $n = 2$. Let $\mathfrak{p}_1, \ldots, \mathfrak{p}_s$ be the finitely many prime divisors in $\operatorname{Ass}(A/\mathfrak{a}_1)$. Choose $x_i \in \mathfrak{p}_i - \mathfrak{p}_1 \cup \cdots \cup \mathfrak{p}_i^2 \cup \cdots \cup \mathfrak{p}_s$. Let $S = A - \mathfrak{p}_1 \cup \cdots \cup \mathfrak{p}_s$, then by 3.31 $S^{-1}A$ is a principal ideal domain. In $S^{-1}A$, $\mathfrak{a}_2 = (x_1^{n_1} \ldots x_s^{n_s})$. By considering a finite set of generators for $\mathfrak{a}_2$ we find a nonzero $x \in K$ such that $x\mathfrak{a}_2 \subseteq A$ and $\mathfrak{p}_i \notin \operatorname{Ass}(A/x\mathfrak{a}_2)$ for any i. We may assume that $\mathfrak{a}_1 + \mathfrak{a}_2 = A$ and conclude by the surjection $\mathfrak{a}_1 \oplus \mathfrak{a}_2 \to A$.

(2) By (1) the projection $M \to M/T$ splits. □

Corollary 3.33. *Let A be a Dedekind domain with fraction field K and let M and M' be a finitely generated torsion free A-modules. Then*

$$M \simeq M' \quad \Leftrightarrow \quad \begin{cases} \mathrm{rank}_K(M \otimes_A K) = \mathrm{rank}_K(M' \otimes_A K) \\ c_1(M) = c_1(M'). \end{cases}$$

Proof. Follows from 3.29 and 3.31. □

Corollary 3.34. *Let A be a principal ideal domain. Then a finitely generated torsion free module is free.*

Proof. Follows from 3.32. □

Chapter 4

Completion

4.1 Exactness of the completion functor

For a projective system $\to M_{n+1} \to M_n \to$ the *projective limit* $\varprojlim M_n$ is the subset of sequences (x_n) in $\prod M_n$ such that $x_{n+1} \mapsto x_n$.

Definition 4.1. Let A denote a noetherian ring and $\mathfrak{I}$ an ideal in A. For an A-module M, consider the projective system

$$\cdots \longrightarrow M/\mathfrak{I}^{n+1}M \longrightarrow M/\mathfrak{I}^n M \longrightarrow \cdots .$$

The projective limit

$$\hat{M} = \varprojlim M/\mathfrak{I}^n M$$

is called the $\mathfrak{I}$-*adic completion* of M. Note that $\hat{M}$ carries a canonical structure as module over the ring $\hat{A} = \varprojlim A/\mathfrak{I}^n$.

Theorem 4.2. *The completion functor $M \mapsto \hat{M}$ is exact on the category of finitely generated A-modules.*

Proof. Consider the exact sequence

$$0 \longrightarrow N \longrightarrow M \longrightarrow P \longrightarrow 0$$

of finitely generated A-modules. By 4.3 below the following two sequences are exact

$$0 \longrightarrow \varprojlim N/\mathfrak{I}^n M \cap N \longrightarrow \varprojlim M/\mathfrak{I}^n M \longrightarrow \varprojlim P/\mathfrak{I}^n P \longrightarrow 0$$
$$0 \longrightarrow \varprojlim \mathfrak{I}^n M \cap N/\mathfrak{I}^n N \longrightarrow \varprojlim N/\mathfrak{I}^n N \longrightarrow \varprojlim N/\mathfrak{I}^n M \cap N \longrightarrow 0.$$

Thus it suffices to prove that

$$\varprojlim \mathfrak{I}^n M \cap N/\mathfrak{I}^n N = 0.$$

By the Artin–Rees lemma we can find $k \in \mathbb{N}$ such that $\mathfrak{J}^n(\mathfrak{J}^k M \cap N) = \mathfrak{J}^{n+k} M \cap N$ and therefore

$$\mathfrak{J}^{n+k} M \cap N / \mathfrak{J}^{n+k} N \to \mathfrak{J}^n M \cap N / \mathfrak{J}^n N$$

is zero for all $n \in \mathbb{N}$.   □

Lemma 4.3. *Let*

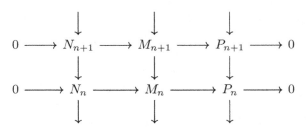

denote a short exact sequence of projective systems of abelian groups indexed by $\mathbb{N}$. *If the projections* $N_{n+1} \to N_n$ *satisfy the* **Mittag–Leffler condition**

$$\mathrm{Im}(N_q \to N_n), \qquad q \gg n$$

stabilize, then

$$0 \longrightarrow \varprojlim N_n \longrightarrow \varprojlim M_n \longrightarrow \varprojlim P_n \longrightarrow 0$$

is exact.

Proof. Straightforward.   □

Corollary 4.4. *Let* M *be a finitely generated A-module, then the canonical map*

$$M \otimes_A \hat{A} \to \hat{M}$$

is an isomorphism.

Proof. This is clear for $M = A^n$. In general choose a presentation

$$L_1 \longrightarrow L_0 \longrightarrow M \longrightarrow 0$$

by finitely generated free A-modules. By 4.2 the following diagram has exact rows

$$
\begin{array}{ccccccc}
L_1 \otimes_A \hat{A} & \longrightarrow & L_0 \otimes_A \hat{A} & \longrightarrow & M \otimes_A \hat{A} & \longrightarrow & 0 \\
\downarrow & & \downarrow & & \downarrow & & \\
\hat{L}_1 & \longrightarrow & \hat{L}_0 & \longrightarrow & \hat{M} & \longrightarrow & 0.
\end{array}
$$

Consequently $M \otimes_A \hat{A} \to \hat{M}$ is an isomorphism.   □

Corollary 4.5. *The canonical map $A \to \hat{A}$ is flat.*

Proof. Follows immediately from 4.2 and 4.4. □

Proposition 4.6. *Let $\mathfrak{I}$ and $\mathfrak{J}$ be ideals in the noetherian ring A. If $V(\mathfrak{I}) = V(\mathfrak{J})$ then for any A-module the $\mathfrak{I}$-adic and the $\mathfrak{J}$-adic completion of M are canonical isomorphic.*

Proof. We can assume $\mathfrak{I} \subseteq \mathfrak{J}$ and $\mathfrak{J}^k \subseteq \mathfrak{I}$ for some $k \in \mathbb{N}$. By 4.3 we have an exact sequence

$$0 \longrightarrow \varprojlim \mathfrak{J}^n M / \mathfrak{I}^n M \longrightarrow \varprojlim M / \mathfrak{I}^n M \longrightarrow \varprojlim M / \mathfrak{J}^n M \longrightarrow 0.$$

On the other hand, the projection

$$\mathfrak{J}^{nk} M / \mathfrak{I}^{nk} M \to \mathfrak{J}^n M / \mathfrak{I}^n M$$

is zero for all $n \in \mathbb{N}$, whence

$$\varprojlim \mathfrak{J}^n M / \mathfrak{I}^n M = 0.$$
□

4.2 Separation of the ℑ-adic topology

Let A denote a noetherian ring and $\mathfrak{I}$ an ideal in A. For an A-module M we can equip M with a structure of topological group, namely that for which $(\mathfrak{I}^n M)$ form a basis for the neighborhoods of 0. This topology is called the $\mathfrak{I}$-*adic topology* on M. From elementary considerations on topological groups follows that M is separated (i.e., Hausdorff) in the $\mathfrak{I}$-adic topology if and only if $\bigcap_{n \in \mathbb{N}} \mathfrak{I}^n M = 0$.

Proposition 4.7. *All finitely generated A-modules are separated in the $\mathfrak{I}$-adic topology if and only if $\mathfrak{I}$ is contained in all maximal ideals of A.*

Proof. Suppose $\mathfrak{I}$ is contained in all maximal ideals of A and let M be a finitely generated A-module. Put $N = \bigcap_{n \in \mathbb{N}} \mathfrak{I}^n M$. By the Artin–Rees lemma we can find $k \in \mathbb{N}$ such that $\mathfrak{I}^{n+k} M \cap N = \mathfrak{I}^n (\mathfrak{I}^k M \cap N)$, from which it follows that $\mathfrak{I}N = N$. For any maximal ideal $\mathfrak{m}$ we get $N_\mathfrak{m} = 0$ by Nakayama's lemma, and therefore $N = 0$. Conversely, let $\mathfrak{m}$ be a maximal ideal, $A/\mathfrak{m}$ is separated in the $\mathfrak{I}$-adic topology if and only if $\mathfrak{I} \subseteq \mathfrak{m}$. □

4.3   Complete filtered rings

A filtered ring $A = (A^n)_{n \in \mathbb{N}}$ is called *complete* if the canonical map

$$A \to \varprojlim A/A^n$$

is an isomorphism. The $\mathfrak{I}$-adic completion is a complete ring.

Proposition 4.8. *Let A be a complete filtered ring. If $\mathrm{gr}(A)$ is noetherian, then A itself is noetherian.*

Proof. Let $\mathfrak{I}$ be an ideal of A. Consider $\mathfrak{I}$ as a filtered A-module by putting $\mathfrak{I}^n = \mathfrak{I} \cap A^n$. The graded $\mathrm{gr}(A)$-module $\mathrm{gr}(\mathfrak{I})$ is finitely generated. Let (x_{n_i}) be a finite set of homogeneous generators for this module. We get a map of filtered modules

$$\bigoplus_i A[-n_i] \to \mathfrak{I}$$

with associated graded map being surjective. Conclusion by the following lemma. $\square$

Lemma 4.9. *Let A be a filtered ring and $\phi : M \to N$ a morphism of filtered A-modules with M complete and N separated. If $\mathrm{gr}(\phi) : \mathrm{gr}(M) \to \mathrm{gr}(N)$ is surjective, then ϕ is surjective.*

Proof. By induction it follows that ϕ induces a surjection $M/M^n \to N/N^n$ for all $n \in \mathbb{Z}$. Consider the commutative diagram:

$$
\begin{array}{ccccccccc}
0 & \longrightarrow & K^{n+1} & \longrightarrow & M/M^{n+1} & \longrightarrow & N/N^{n+1} & \longrightarrow & 0 \\
 & & \downarrow & & \downarrow & & \downarrow & & \\
0 & \longrightarrow & K^n & \longrightarrow & M/M^n & \longrightarrow & N/N^n & \longrightarrow & 0 \\
 & & & & \downarrow & & \downarrow & & \\
 & & & & 0 & & 0 & &
\end{array}
$$

From the fact that $\mathrm{gr}(\phi)$ is surjective one deduces by means of the snake lemma that $K^{n+1} \to K^n$ is surjective. Using 4.3 we conclude that ϕ induces a surjective map $\varprojlim M/M^n \to \varprojlim N/N^n$. From this follows that $N \simeq \varprojlim N/N^n$ and therefore $\phi : M \to N$ is surjective. $\square$

Proposition 4.10. *Let $\mathfrak{I}$ be an ideal in the noetherian ring A. Then the $\mathfrak{I}$-adic completion $\hat{A}$ of A is a noetherian ring.*

Proof. Put $\hat{\mathfrak{J}} = \varprojlim \mathfrak{J}/\mathfrak{J}^n$ and consider the $\hat{\mathfrak{J}}$-adic filtration on $\hat{A}$. We have $\mathrm{gr}(A) \simeq \mathrm{gr}(\hat{A})$ by 4.2 and 4.4. The ring $\mathrm{gr}(A)$ is noetherian by Hilbert's basis theorem. Conclusion by 4.8. $\qquad\square$

Example 4.11. Let A be a noetherian ring. Then the ring of formal power series

$$A[[T_1, \ldots, T_n]]$$

is a noetherian ring.

Proposition 4.12. *Let $\mathfrak{J}$ be an ideal in the ring A. If A is complete in the $\mathfrak{J}$-adic topology, then any maximal ideal of A contains $\mathfrak{J}$.*

Proof. Let us first prove that $1 - a$ is invertible for any $a \in \mathfrak{J}$. For this consider the residue class of $\sum_{i=0}^{n-1} a^i \mod \mathfrak{J}^n$. It follows that we can find $b \in A$ such that

$$b = \sum_{i=0}^{n-1} a^i \mod \mathfrak{J}^n, \qquad \text{for all } n \in \mathbb{N}.$$

Clearly $(1 - a)b = 1$. Next given a maximal ideal $\mathfrak{m}$ in A and $a \in \mathfrak{J}$. Suppose $a \neq 0 \mod \mathfrak{m}$. Then we can find $c \in A$ such that $1 - ca = 0 \mod \mathfrak{m}$. On the other hand, $ca \in \mathfrak{J}$, so $1 - ca$ is invertible, giving a contradiction. $\qquad\square$

Corollary 4.13. *Let $\mathfrak{J}$ be an ideal in the noetherian ring A. Then $\mathfrak{m} \mapsto \hat{\mathfrak{m}}$ induces a bijective correspondence between the set of maximal ideals in A containing $\mathfrak{J}$ and the set of all maximal ideals in the $\mathfrak{J}$-adic completion $\hat{A}$.*

Proof. Follows from 4.5 and 4.12. $\qquad\square$

4.4 Completion of local rings

Let A denote a noetherian local ring with maximal ideal $\mathfrak{m}$ and residue field k. The $\mathfrak{m}$-adic completion of A will be called the *completion* of A and will be denoted $\hat{A}$. Let us recall 4.10 and 4.12 that $\hat{A}$ is a noetherian local ring with maximal ideal $\hat{\mathfrak{m}} = \mathfrak{m}\hat{A}$ and that

$$\mathrm{gr}_{\mathfrak{m}}(A) \simeq \mathrm{gr}_{\mathfrak{m}}(\hat{A}).$$

Proposition 4.14.

(1) $\dim(A) = \dim(\hat{A})$.

(2) A *is regular if and only if* $\hat{A}$ *is regular.*

(3) A *is Cohen–Macaulay if and only if* $\hat{A}$ *is Cohen–Macaulay.*

Proof. 1.41, 1.48 and 4.15 below.   □

Proposition 4.15. *Let* $M \neq 0$ *be a finitely generated A-module and* $\hat{M} = M \otimes_A \hat{A}$. *Then*

(1) $\dim(M) = \dim(\hat{M})$.

(2) $\operatorname{proj\,dim}(M) = \operatorname{proj\,dim}(\hat{M})$.

(3) $\operatorname{depth}(M) = \operatorname{depth}(\hat{M})$.

Proof. (1) Let us remark that $\operatorname{gr}_{\mathfrak{m}}(M) = \operatorname{gr}_{\hat{\mathfrak{m}}}(\hat{M})$ and whence $\chi_{\mathfrak{m}}(M, T) = \chi_{\hat{\mathfrak{m}}}(\hat{M}, T)$. Thus it suffices to recall that the dimension is given by, 2.18,

$$\deg \chi_{\mathfrak{m}}(M, T) = \dim(M).$$

(2) By 4.16 below we have

$$\operatorname{Tor}_i^A(k, M) \simeq \operatorname{Tor}_i^{\hat{A}}(k, \hat{M})$$

since $\operatorname{Tor}^A(k, M)$ is annihilated by $\mathfrak{m}$.

(3) Let us prove that

$$\operatorname{Ext}_A^i(k, M) \simeq \operatorname{Ext}_{\hat{A}}^i(k, \hat{M}).$$

By 4.16

$$\operatorname{Ext}_{\hat{A}}^i(k, \hat{M}) = \operatorname{Ext}_A^i(k, M) \otimes_A \hat{A}.$$

Conclusion by the fact that $\operatorname{Ext}_A(k, M)$ is annihilated by $\mathfrak{m}$.   □

Lemma 4.16. *Let M and N be a finitely generated A-module. Then*

$$\operatorname{Tor}_i^A(N, M) \otimes_A \hat{A} \simeq \operatorname{Tor}_i^{\hat{A}}(\hat{N}, \hat{M}),$$
$$\operatorname{Ext}_A^i(N, M) \otimes_A \hat{A} \simeq \operatorname{Ext}_{\hat{A}}^i(\hat{N}, \hat{M}).$$

Proof. Let $L.$ be a resolution of M by finitely generated free A-modules. We have, since $A \to \hat{A}$ is flat that

$$\mathrm{Tor}_i^{\hat{A}}(\hat{N}, \hat{M}) = H_i(\hat{N} \otimes_{\hat{A}} (\hat{A} \otimes_A L.)) = H_i(N \otimes_A L.) \otimes_A \hat{A} = \mathrm{Tor}_i^A(k, M) \otimes_A \hat{A}.$$

Choose a resolution $K.$ of N by finitely generated free A-modules. We have

$$\mathrm{Hom}_{\hat{A}}(K. \otimes_A \hat{A}, \hat{M}) = \mathrm{Hom}_A(K., M) \otimes_A \hat{A}$$

and whence

$$\mathrm{Ext}_{\hat{A}}^i(\hat{N}, \hat{M}) = \mathrm{Ext}_A^i(N, M) \otimes_A \hat{A}. \qquad \square$$

4.5 Structure of complete local rings

The following theorem is due to I.S. Cohen.

Theorem 4.17 (Cohen's structure theorem). *Any complete noetherian local ring is a quotient of a complete regular local ring.*

Proof. The proof will not be given here, see Grothendieck, *Eléments de géométrie algébrique*, Inst. Hautes Études Sci. Pub. Math. 20 (1964), section IV.0.19.8. $\qquad \square$

Chapter 5

Injective Modules

5.1 Injective modules

Let A denote a ring, an A-module E is called *injective* if the functor $\mathrm{Hom}_A(-, E)$ is exact.

Proposition 5.1. *The A-module E is injective if and only if for every ideal $\mathfrak{I}$, any linear map $\mathfrak{I} \to E$ extends to a linear map $A \to E$.*

Proof. To prove that E is injective it suffices to prove that whenever $i : F \to F'$ is an injection of A-modules, then any linear map $f : F \to E$ extends to a linear map $F' \to E$. Consider the push-out diagram, $E' = \mathrm{Cok}((i, -f) : F \to F' \oplus E)$,

$$
\begin{array}{ccc}
F & \xrightarrow{\;i\;} & F' \\
\downarrow{f} & & \downarrow{f'} \\
E & \xrightarrow{\;j\;} & E'
\end{array}
$$

and note that j is injective. Thus it suffices to prove that j has a retraction. Consider the set of pairs (G, r) where G is a submodule of E' containing $\mathrm{Im}(j)$ and $r : G \to E$ is a retraction to $j : E \to G$. Order this set such that $(G', r') \leq (G'', r'')$ means that $G' \subseteq G''$ and r'' is an extension of r'. Use Zorn's lemma to choose a maximal element in this set, say, (E'', ρ). We are going to prove that $E'' = E'$. Suppose $E'' \neq E'$ and choose $e \in E' - E''$. Put $\mathfrak{I}_e = \{a \in A \mid ae \in E''\}$. The canonical diagram

$$
\begin{array}{ccc}
\mathfrak{I}_e & \longrightarrow & A \\
\downarrow & & \downarrow \\
E'' & \longrightarrow & E'' + Ae
\end{array}
$$

is clearly a push-out diagram. Finally extend the composite of $\mathfrak{J}_e \to E''$ to all of A and obtain an extension of ρ to $E'' + Ae$, contradicting the maximality of E''. $\qquad\square$

Corollary 5.2. *Let A be a principal ideal domain. An A-module E is injective if and only if E is divisible (i.e., for $0 \neq a \in A$ the map $E \to E$, $x \to ax$ is surjective).*

Proof. To extend $f : (a) \to E$, choose $e \in E$ such that $ae = f(a)$ and define $A \to E$, $1 \to e$. $\qquad\square$

Proposition 5.3. *Any A-module admits an injection into an injective A-module.*

Proof. Remark that the abelian group $\mathbb{Q}/\mathbb{Z}$ is injective, and that for any abelian group M, the evaluation map

$$ev : M \to \mathrm{Hom}_{\mathbb{Z}}(\mathrm{Hom}_{\mathbb{Z}}(M, \mathbb{Q}/\mathbb{Z}), \mathbb{Q}/\mathbb{Z})$$

is injective. Suppose that M is an A-module. Choose a free A-module L and an A-linear surjection

$$L \to \mathrm{Hom}_{\mathbb{Z}}(M, \mathbb{Q}/\mathbb{Z}).$$

Using the evaluation map above, this embeds M into the A-module

$$\mathrm{Hom}_{\mathbb{Z}}(L, \mathbb{Q}/\mathbb{Z}).$$

We are going to prove that this module is an injective A-module. Remark first that this module is a product of modules of the form

$$\mathrm{Hom}_{\mathbb{Z}}(A, \mathbb{Q}/\mathbb{Z}),$$

since an arbitrary product of injective modules is injective it suffices to prove that $\mathrm{Hom}_{\mathbb{Z}}(A, \mathbb{Q}/\mathbb{Z})$ is an injective A-module. Remark that for any A-module N, we have

$$\mathrm{Hom}_A(N, \mathrm{Hom}_{\mathbb{Z}}(A, \mathbb{Q}/\mathbb{Z})) \simeq \mathrm{Hom}_{\mathbb{Z}}(N, \mathbb{Q}/\mathbb{Z}). \qquad\square$$

Definition 5.4. Given an A-module M, then there is a resolution

$$0 \longrightarrow M \xrightarrow{\partial^{-1}} E^0 \xrightarrow{\partial^0} E^1 \longrightarrow \cdots \longrightarrow E^i \xrightarrow{\partial^i} E^{i+1} \longrightarrow \cdots$$

by injective modules. The length of a shortest injective resolution is called the *injective dimension* and is denoted $\mathrm{inj\,dim}(M)$.

5.2 Injective envelopes

Let A denote a ring. A linear injection $N \to M$ of A-modules is said to be an *essential extension* if any nonzero submodule of M has a nonzero intersection with N. Note that the composite of two essential extensions is an essential extension and that a direct sum of any family of essential extensions is an essential extension.

Definition 5.5. Given an A-module M. An *injective envelope* of M is an essential extension $M \to E$ with E injective.

Theorem 5.6. *Any A-module M has an injective envelope. Any two injective envelopes are isomorphic as extensions of M.*

Proof. The uniqueness of injective envelopes is straightforward. For the existence choose an extension $M \to E'$ with E' injective, 5.3. Let E be a submodule of E' containing M and such that $M \to E$ is an essential extension. Moreover choose E maximal among such extensions which have these two properties by Zorn's lemma. As is easily seen the module E has the property that any essential extension of E is trivial. We shall prove that this property of E implies that E is injective. For this let $E \to F$ be any extension of E, We shall prove that it has a retraction. Let N be a submodule of F with $N \cap F = 0$ and maximal with this property by Zorn's lemma. Note that the composite $E \to F \to F/N$ is an essential extension of E, thus this composite is an isomorphism. $\square$

Corollary 5.7. *Let E be an indecomposable (having no nontrivial direct summands) injective module. Any two nonzero submodules of E have a nonzero intersection.*

Proof. Let $N \neq 0$ be a submodule of E and $N \to E(N)$ an injective envelope of N. The embedding $N \to E$ extends to give a linear map $E(N) \to E$. This map is an injection since $N \to E(N)$ is an essential extension. Since $E(N)$ is injective, $E(N)$ is a nonzero direct summand of E, whence $E \simeq E(N)$. $\square$

Definition 5.8. Let $M \to E$ be an injective envelope of the module M. The module E (which is unique up to isomorphism) is called an *injective hull* of M and is denoted $E_A(M)$ or just $E(M)$.

Proposition 5.9. *Let* $\mathfrak{p}$ *be a prime ideal in the ring* A*. Then*

(1) $E(A/\mathfrak{p})$ *is indecomposable.*

(2) $\mathfrak{p}$ *is determined by* $E(A/\mathfrak{p})$*. That is, if* $\mathfrak{q}$ *is a prime ideal and* $E(A/\mathfrak{p}) \simeq E(A/\mathfrak{q})$ *then* $\mathfrak{p} = \mathfrak{q}$*.*

(3) *If* A *is noetherian then any indecomposable injective module is of the form* $E(A/\mathfrak{p})$ *for some prime ideal* $\mathfrak{p}$*.*

Proof. (1) and (2) follows immediately from 5.6 and 5.7. (3) follows from $\mathrm{Ass}(E) \neq \emptyset$ and 5.7. $\qquad \square$

5.3   Decomposition of injective modules

Throughout this section A denotes a noetherian ring.

Proposition 5.10. *A direct sum of any family of injective* A*-modules is an injective* A*-module.*

Proof. Follows from 5.1 since any ideal is finitely generated. $\qquad \square$

Theorem 5.11. *An injective* A*-module is the direct sum of modules of the form* $E(A/\mathfrak{p})$ *where* $\mathfrak{p}$ *is a prime ideal.*

Proof. Let $E \neq 0$ be an injective A-module. Note $\mathrm{Ass}(E) \neq \emptyset$. Consider subsets of $T \subseteq E$ such that (a) the annihilator of any element of T is a prime ideal, and (b) the submodule generated by T is the direct sum of the submodules $(At)_{t \in T}$. Choose T maximal among subsets with properties (a) and (b) by Zorn's lemma. Let E_T denote the injective envelope of the submodule generated by T, we may consider E_T as a direct summand of E. If $E_T \neq E$ we can choose a complement in E, say, Q. Again since $Q \neq 0$ we can find a $q \in Q$ whose annihilator is a prime ideal, contradicting the maximality of T. Whence $E \simeq \oplus_t E(At)$ as it follows from 5.10. $\qquad \square$

Remark 5.12. We shall prove below in 5.15 that the decomposition in 5.11 is unique in the following sense. Let $(E_i)_{i \in I}$ and $(F_j)_{i \in J}$ be two families of indecomposable injective modules. If

$$\bigoplus_{i \in I} E_i \simeq \bigoplus_{j \in J} F_j,$$

then there exists a bijection $\theta : I \to J$ such that

$$E_i \simeq F_{\theta(i)}.$$

Proposition 5.13. *Let S be a multiplicative subset. The localization functor*

$$M \mapsto S^{-1}M$$

transforms an injective A-module into an injective $S^{-1}A$-module and an essential extension of A-modules into an essential extension of $S^{-1}A$-modules.

Proof. The last part is left to the reader.

Note first that any injection in the category of finitely generated $S^{-1}A$-modules is of the form $S^{-1}f : S^{-1}M \to S^{-1}N$ where $f : M \to N$ is an injection of finitely generated A-modules. Next note that for a finitely generated A-module N and any A-module P, the canonical map

$$S^{-1} \operatorname{Hom}_A(N, P) \to \operatorname{Hom}(S^{-1}N, S^{-1}P)$$

is an isomorphism, as it follows by considering a presentation $A^m \to A^n \to N \to 0$. Finally apply 5.1. $\qquad\square$

Corollary 5.14. *Let S be a multiplicative subset of A and $\mathfrak{p}$ a prime ideal in A. Then*

$$S^{-1}E(A/\mathfrak{p}) \simeq \begin{cases} 0 & \text{if } \mathfrak{p} \cap S \neq \emptyset, \\ E_{S^{-1}A}(S^{-1}A/S^{-1}\mathfrak{p}) & \text{if } \mathfrak{p} \cap S = \emptyset. \end{cases}$$

Moreover in the last case, the canonical map $E(A/\mathfrak{p}) \to S^{-1}E(A/\mathfrak{p})$ is an isomorphism.

Proof. $S^{-1}(A/\mathfrak{p}) \to S^{-1}E(A/\mathfrak{p})$ is essential, giving the first part. For the second part note that any $s \in S$ is a nozero divisor on the indecomposable injective module $E(A/\mathfrak{p})$ and therefore multiplication by s is an isomorphism. $\qquad\square$

Corollary 5.15. *Let $\mathfrak{p}$ be a prime ideal in A and $E = \oplus_{i \in I} E_i$ a decomposition of an injective module into indecomposables. Then*

$$\text{Cardinality}\{i \in I \mid E_i \simeq E(A/\mathfrak{p})\} = \operatorname{rank}_{k_\mathfrak{p}} \operatorname{Hom}_{A_\mathfrak{p}}(k_\mathfrak{p}, E_\mathfrak{p}).$$

Proof. By 5.14 we may assume A local with maximal ideal $\mathfrak{p}$. Note

$$\operatorname{Hom}_A(k, E(A/\mathfrak{q})) = \begin{cases} k & \text{if } \mathfrak{q} = \mathfrak{p} \\ 0 & \text{if } \mathfrak{q} \neq \mathfrak{p}, \end{cases}$$

from which the conclusion follows. $\qquad\square$

5.4 Matlis duality

Throughout this section A denotes a noetherian local ring with maximal ideal $\mathfrak{m}$ and residue field k. The $\mathfrak{m}$-adic completion of A will be denoted by $\hat{A}$.

Definition 5.16. A module is called *artinian* if any decreasing sequence of submodules is stationary or equivalently any nonempty subset of submodules of M contains a minimal element. Let $0 \to N \to M \to P \to 0$ be an exact sequence. Then M is artinian if and only if N and P are artinian.

Proposition 5.17. *The functor which to an $\hat{A}$-module assigns the underlying A-module induces an equivalence between the category of artinian $\hat{A}$ modules and the category of artinian A-modules. Moreover this equivalence transforms injective $\hat{A}$-modules to injective A-modules and conversely, an artinian $\hat{A}$-module is injective if the underlying A-module is injective.*

Proof. Let us call a module primary if any finitely generated submodule has finite length. Clearly any quotient module and submodule of a primary module is again primary. For the first part of the proposition it suffices to show that the forgetful functor induces an equivalence from the category of primary $\hat{A}$-modules to that of primary A-modules. We shall construct an inverse functor. Let D be a primary A-module. Put

$$D_i = \{d \in D \mid \mathfrak{m}^i d = 0\}.$$

Note that

$$D = \bigcup_{i \in \mathbb{N}} D_i$$

$$\operatorname{End}_A(D) = \varprojlim \operatorname{Hom}_A(D_i, D) = \varprojlim \operatorname{End}_A(D_i).$$

For each i consider the commutative diagram

$$
\begin{array}{ccc}
A/\mathfrak{m}^{i+1} & \longrightarrow & \operatorname{End}_A(D_{i+1}) \\
\downarrow & & \downarrow \\
A/\mathfrak{m}^i & \longrightarrow & \operatorname{End}_A(D_i).
\end{array}
$$

This defines an A-linear map $\hat{A} \to \operatorname{End}_A(D)$, i.e., an $\hat{A}$-module structure on D. The rest of the details are left to the reader. To prove the last part it will suffice to prove that if an artinian module E is an injective object in the category of artinian modules, then E is an injective module. So, consider the diagram

$$
\begin{array}{ccc}
0 \longrightarrow N \longrightarrow M \\
\downarrow{\scriptstyle \phi} \\
E
\end{array}
$$

with M and N finitely generated. Replacing M by $M/\operatorname{Ker}(\phi)$ and N by $N/\operatorname{Ker}(\phi)$ we can assume that N has finite length. Choose a submodule Q of M with $\mathfrak{m} \notin \operatorname{Ass}(Q)$ and maximal with that property. An easy argument shows that $\operatorname{Ass}(M/Q) = \{\mathfrak{m}\}$, i.e., M/Q has finite length. Clearly $Q \cap N = 0$. We can extend ϕ to M/Q. $\qquad\square$

Proposition 5.18. *The injective envelope of an artinian module is artinian.*

Proof. Let D be the injective envelope of an artinian module. It follows easily that D is the direct sum of finitely many copies of $E = E(k)$, the injective envelope of the residue field. To show that E is artinian we may assume that A is complete. By proposition 5.19 below we have $A = \operatorname{End}(E)$. For a submodule F of E put

$$
\mathfrak{I}(F) = \operatorname{Ker}(\operatorname{End}_A(E) \to \operatorname{Hom}_A(F, E)).
$$

It suffices to prove if $F_2 \subset F_1 \subseteq E$ then $\mathfrak{I}(F_1) \subset \mathfrak{I}(F_2)$. Consider the exact commutative diagram

$$
\begin{array}{ccccccc}
0 & \longrightarrow & \operatorname{Hom}_A(E, E) & \longrightarrow & \operatorname{Hom}_A(E, E) & \longrightarrow & 0 \\
& & \downarrow & & \downarrow & & \downarrow \\
0 \longrightarrow \operatorname{Hom}_A(F_1/F_2, E) & \longrightarrow & \operatorname{Hom}_A(F_1, E) & \longrightarrow & \operatorname{Hom}_A(F_2, E)
\end{array}
$$

to establish a short exact sequence

$$
0 \longrightarrow \mathfrak{I}(F_1) \longrightarrow \mathfrak{I}(F_2) \longrightarrow \operatorname{Hom}_A(F_1/F_2, E) \longrightarrow 0.
$$

We have $\operatorname{Ass}(F_1/F_2) = \{\mathfrak{m}\}$ from which we deduce $\operatorname{Hom}_A(k, F_1/F_2) \neq 0$ and therefore $\operatorname{Hom}_A(F_1/F_2, E) \neq 0$. $\qquad\square$

Proposition 5.19. *Let A be a noetherian local ring and E the injective envelope of the residue field. Then*

$$
\hat{A} = \operatorname{End}_A(E).
$$

Proof. Note that $\mathrm{Hom}_A(k, E) \simeq k$. Deduce from this by induction on the length that for any A-module of finite length N, the evaluation map

$$N \to \mathrm{Hom}_A(\mathrm{Hom}_A(N, E), E)$$

is an isomorphism. With the notation introduced in the proof of 5.17 it follows that the canonical map $A/\mathfrak{m}^i \to \mathrm{End}_A(E_i)$ is an isomorphism. Whence

$$\hat{A} = \varprojlim \mathrm{End}_A(E_i) = \mathrm{End}_A(E). \qquad \square$$

We shall now show that for a complete ring, the category of noetherian modules is dual to the category of artinian modules.

Theorem 5.20 (Matlis duality). *Let A be a complete noetherian local ring and E the injective envelope of the residue field. If M is a noetherian/artinian A-module, then the Matlis dual $M^\vee = \mathrm{Hom}_A(M, E)$ is artinian/noetherian. If M is either noetherian or artinian then the evaluation map*

$$M \to \mathrm{Hom}_A(\mathrm{Hom}_A(M, E), E) = M^{\vee\vee}$$

is an isomorphism.

Proof. For M choose a presentation $A^n \to A^m \to M \to 0$ resp. a copresentation $0 \to M \to E^m \to E^n$ in order to reduce to the case $M = A$ resp. $M = E$. Conclusion by 5.18 and 5.19. $\qquad \square$

Corollary 5.21. *Let A be a noetherian local ring and E the injective envelope of the residue field. If M is an artinian A-module, then the composition map*

$$\mathrm{Hom}_A(E, M) \to \mathrm{Hom}_A(\mathrm{Hom}_A(M, E), \hat{A})$$

is an isomorphism.

Proof. Straightforward using 5.19 and 5.20. $\qquad \square$

Corollary 5.22. *Let A be an Artin local ring. Then any artinian module has finite length. The A-module A is injective if and only if $\mathrm{Hom}_A(k, A) = k$.*

Proof. The category of artinian modules is dual to the category of modules of finite length, whence the first conclusion. Under this duality, A corresponds to E, whence $\ell_A(A) = \ell_A(E)$. Note also that A is an essential extension of $\mathrm{Hom}_A(k, A)$. Whence

$$\ell_A(E(A)) = \ell_A(A)\,\ell_A(\mathrm{Hom}_A(k, A)).$$

In particular $E(A) = A$ if and only if $\mathrm{Hom}_A(k, A) \simeq k$. $\qquad \square$

5.5 Minimal injective resolutions

Let A denote a noetherian ring.

Definition 5.23. Let M be an A-module. An injective resolution of M

$$0 \longrightarrow M \xrightarrow{\partial^{-1}} E^0 \xrightarrow{\partial^0} E^1 \longrightarrow \cdots \longrightarrow E^i \xrightarrow{\partial^i} E^{i+1} \longrightarrow \cdots$$

is called a *minimal* injective resolution if $\operatorname{Im}(\partial^{i-1}) \to E^i$ is an injective envelope for all $i \in \mathbb{N}$. The length of a minimal injective resolution is the injective dimension.

Proposition 5.24. *Let M be a finitely generated A-module and $E^{\cdot}(M)$ a minimal injective resolution of M. For $i \in \mathbb{N}$ we have*

$$E^i(M) \simeq \bigoplus_{\mathfrak{p} \in \operatorname{Spec}(A)} E(A/\mathfrak{p})^{\mu^i(\mathfrak{p}, M)}$$

where

$$\mu^i(\mathfrak{p}, M) = \operatorname{rank}_{k_\mathfrak{p}} \operatorname{Ext}^i_{A_\mathfrak{p}}(k_\mathfrak{p}, M_\mathfrak{p}).$$

Proof. By the results of section 5.3 we may assume that A is local with maximal ideal $\mathfrak{m}$ and residue field k. Note first that the differential in the complex $\operatorname{Hom}_A(k, E^{\cdot}(M))$ is zero. Next notice that

$$\operatorname{Hom}_A(k, E(A/\mathfrak{p})) = \begin{cases} 0 & \mathfrak{p} \neq \mathfrak{m} \\ k & \mathfrak{p} = \mathfrak{m}. \end{cases}$$

From this the conclusion follows easily. $\qquad\square$

Proposition 5.25. *Let $\mathfrak{p} \subset \mathfrak{q}$ be prime ideals in A with no prime ideal lying properly between $\mathfrak{p}$ and $\mathfrak{q}$. Then for any finitely generated A-module M we have (with the notation of 5.24) that*

$$\mu^i(\mathfrak{p}, M) \neq 0 \quad \Rightarrow \quad \mu^{i+1}(\mathfrak{q}, M) \neq 0.$$

Proof. We may assume that A is local with maximal ideal $\mathfrak{q}$ and residue field k. Choose $a \in \mathfrak{q} - \mathfrak{p}$ and consider the exact sequence

$$0 \longrightarrow A/\mathfrak{p} \xrightarrow{a} A/\mathfrak{p} \longrightarrow Q \longrightarrow 0.$$

Assume $\operatorname{Ext}^{i+1}_A(k, M) = 0$, then $\operatorname{Ext}^{i+1}_A(R, M) = 0$ for any module R of finite length, so we get an exact sequence

$$\operatorname{Ext}^i_A(A/\mathfrak{p}, M) \xrightarrow{a} \operatorname{Ext}^i_A(A/\mathfrak{p}, M) \longrightarrow 0.$$

Thus by Nakayama's lemma $\mathrm{Ext}_A^i(A/\mathfrak{p}, M) = 0$ and whence we have $\mu^i(A/\mathfrak{p}, M) = 0$. □

Corollary 5.26. *Let A be a noetherian local ring with maximal ideal $\mathfrak{m}$ and residue field k and consider a finitely generated module M. Then*

$$\mathrm{inj\,dim}(M) \le d \Leftrightarrow \mathrm{Ext}_A^i(k, M) = 0, \qquad \text{for all } i > d.$$

Proof. If $\mathrm{Ext}_A^i(k, M) = 0$ then $\mu^i(\mathfrak{m}, M) = 0$. By 5.25 $\mu^{i-1}(\mathfrak{p}, M) = 0$ for $\mathfrak{p} \ne \mathfrak{m}$ and $i > d$, so the length of a minimal injective resolution is at most d. □

Corollary 5.27. *Let A be a noetherian local ring with residue field k. If the finitely generated module M has the property*

$$\mathrm{Ext}_A^i(k, M) = 0, \qquad \text{for } i \gg 0,$$

then M has a finite injective resolution.

Proof. By 5.25 $\mu^i(\mathfrak{p}, M) = 0$ for $i \gg 0$. □

5.6   Modules of finite injective dimension

Let A denote a noetherian local ring with maximal ideal $\mathfrak{m}$ and residue field k.

Proposition 5.28. *Let $M \ne 0$ be a finitely generated A-module. If M has a finite injective resolution, then the length of a minimal injective resolution is $\mathrm{depth}(A)$, i.e.,*

$$\mathrm{inj\,dim}(M) = \mathrm{depth}(A).$$

Proof. Let $0 \to M \to E^{\bullet}$ be a minimal injective resolution of M and n the length of $E^{\bullet}$. By 5.24 and 5.25 we have $\mathrm{Ext}_A^n(k, M) \ne 0$, and the functor $\mathrm{Ext}_A^n(-, M)$ is left exact. In particular $\mathrm{Ext}_A^n(D, M) \ne 0$ for any finitely generated A-module $D \ne 0$ with $\mathrm{depth}(D) = 0$. Let $d = \mathrm{depth}(A)$ and choose a sequence $a_1, \dots, a_d \in \mathfrak{m}$ where a_i is a nonzero divisor for $A/(a_1, \dots, a_{i-1})$, $i = 1, \dots, d$. The Koszul complex $K_{\bullet}(a_1, \dots, a_d)$ yields a free resolution of $A/(a_1, \dots, a_d)$ of length d. Since

$$\mathrm{Ext}_A^n(A/(a_1, \dots, a_d), M) \ne 0$$

we get $n \le d$.

Choose a surjection $M \to k \to 0$ and note that we have an exact sequence

$$\operatorname{Ext}_A^d(A/(a_1,\ldots,a_d),M) \longrightarrow \operatorname{Ext}_A^d(A/(a_1,\ldots,a_d),k) \longrightarrow 0.$$

By explicit calculation we have

$$\operatorname{Ext}_A^d(A/(a_1,\ldots,a_d),k) \simeq k.$$

Whence

$$\operatorname{Ext}_A^d(A/(a_1,\ldots,a_d),M) \neq 0$$

and therefore $d \leq n$. $\qquad\square$

Corollary 5.29. *Let M be a finitely generated A-module of finite injective dimension, then*

$$\dim(A/\mathfrak{p}) \leq \operatorname{depth}(A)$$

for all $\mathfrak{p} \in \operatorname{Supp}(M)$.

Proof. Let $\mathfrak{p}_0 \subset \cdots \subset \mathfrak{p}_d = \mathfrak{m}$ be a maximal chain in $\operatorname{Supp}(M)$. By 5.25

$$\mu^0(\mathfrak{p}_0,M) \neq 0 \quad \Rightarrow \quad \mu^d(\mathfrak{m},M) \neq 0.$$

The inequality follows from 5.28. $\qquad\square$

More precisely we have the following proposition.

Proposition 5.30. *Let M be a finitely generated A-module of finite injective dimension, then*

$$\dim(A/\mathfrak{p}) + \operatorname{depth}(A_\mathfrak{p}) = \operatorname{depth}(A)$$

for all $\mathfrak{p} \in \operatorname{Supp}(M)$.

Proof. Follows from 5.29 and 5.32 below. $\qquad\square$

Proposition 5.31. *Let $M \neq 0$ be a finitely generated A-module of finite injective dimension, then*

$$\dim(M) + \inf\{i \mid \operatorname{Ext}_A^i(M,A) \neq 0\} = \operatorname{depth}(A).$$

Proof. Follows from 5.30 and 5.32 below. $\qquad\square$

Lemma 5.32. *Let $M \neq 0$ be a finitely generated A-module, then*

$$\operatorname{depth}(A) \leq \dim(M) + \inf\{i \mid \operatorname{Ext}_A^i(M,A) \neq 0\} \leq \dim(A).$$

Proof. If $\mathrm{depth}(A) = 0$ then $\mathrm{Hom}_A(M, A) \neq 0$ and we are done. Otherwise choose a nonzero divisor $x \in \mathfrak{m}$ not contained in the minimal primes in $\mathrm{Supp}(M)$. By induction on $\dim(A)$ the claim follows from considering the ring $A/(x)$. $\qquad\square$

Proposition 5.33. *Let $M \neq 0$ be a finitely generated A-module of finite injective dimension. Then for any finitely generated A-module N we have*

$$\mathrm{depth}(N) + \sup\{i \mid \mathrm{Ext}_A^i(N, M) \neq 0\} = \mathrm{depth}(A).$$

Proof. By 4.15 we may assume that A is complete. We let E denote the injective envelope of the residue field and by Matlis duality we calculate

$$\mathrm{Ext}_A^i(N, M) \simeq \mathrm{Hom}_A(\mathrm{Tor}_i^A(N, \mathrm{Hom}_A(M, E)), E). \qquad (5.1)$$

Put $\mathrm{depth}(A) = d$. If $\mathrm{depth}(N) = 0$ then by 5.28 the injection $0 \to k \to N$ gives the exact sequence

$$\mathrm{Ext}_A^d(N, M) \longrightarrow \mathrm{Ext}_A^d(k, M) \longrightarrow 0.$$

Conclusion by 5.26. Otherwise let $\mathrm{depth}(N) = n > 0$ and choose a nonzero divisor on N, $x \in \mathfrak{m}$ and consider $0 \to N \to N \to N/xN \to 0$. By induction on n

$$\sup\{i \mid \mathrm{Ext}_A^i(N/xN, M) \neq 0\} = d - n + 1.$$

So by the exact sequence

$$\mathrm{Ext}_A^{d-n}(N, M) \longrightarrow \mathrm{Ext}_A^{d-n}(N, M) \longrightarrow \mathrm{Ext}_A^{d-n+1}(N/xN, M)$$
$$\longrightarrow \mathrm{Ext}_A^{d-n+1}(N, M)$$

it suffices to prove

$$\mathrm{Ext}_A^i(N, M) = 0, \qquad i > d - n.$$

By (5.1) we have

$$\mathrm{Hom}_A(\mathrm{Tor}_i^A(N/xN, \mathrm{Hom}_A(M, E)), E) = 0, \qquad i > d - n + 1$$

giving the exact sequence

$$0 \longrightarrow \mathrm{Tor}_i^A(N, \mathrm{Hom}_A(M, E)) \longrightarrow \mathrm{Tor}_i^A(N, \mathrm{Hom}_A(M, E)),$$

for $i \geq d - n + 1$. Using that $\mathrm{Hom}_A(M, E)$ is artinian and (5.1) we are done. $\square$

Example 5.34. Let A be a Cohen–Macaulay local ring of dimension d, and $a_1, \ldots, a_d$ elements of $\mathfrak{m}$ generating an ideal of finite colength. Consider

$$E = E_{A/(a_1, \ldots, a_d)}(k)$$

as an A-module. It follows from 5.22 that E has finite length. We are going to prove that the A-module E has a finite injective resolution. Consider the functor $\mathrm{Hom}_A(k, -)$ as composed of the functors

$$\{A - mod\} \xrightarrow{\mathrm{Hom}_A(A/(a.), -)} \{A/(a_1 \ldots, a_d) - mod\}$$

$$\xrightarrow{\mathrm{Hom}_{A/(a.)}(k, -)} \{k - mod\}.$$

Remark that the first functor transforms injective A-modules into injective $A/(a_1, \ldots, a_d)$-modules. Using the Koszul complex we get for any A-modules D annihilated by $(a_1, \ldots, a_d)$ that

$$\mathrm{Ext}_A^i(A/(a_1, \ldots, a_d), D) \simeq D^{\binom{d}{i}}$$

in particular

$$\mathrm{Ext}_A^i(A/(a_1, \ldots, a_d), E)$$

is an injective $A/(a_1, \ldots, a_d)$-module, which is zero for $i > d$. Consequently

$$\mathrm{Ext}_A^d(k, A) \simeq \mathrm{Hom}_A(k, A/(a_1, \ldots, a_d))$$

and

$$\mathrm{rank}_k \, \mathrm{Ext}_A^i(k, E) = \binom{d}{i} \mathrm{rank}_k \, \mathrm{Ext}_A^d(k, A).$$

5.7 Gorenstein rings

Let A denote a noetherian local ring with maximal ideal $\mathfrak{m}$ and residue field k.

Proposition 5.35. *The A-module A has a finite injective resolution if and only if A is a Cohen–Macaulay ring of dimension d and*

$$\mathrm{Ext}_A^d(k, A) \simeq k.$$

Proof. In virtue of 5.29 we may assume that A is Cohen–Macaulay and in virtue of 5.36 below we may assume that A is an Artin local ring. Conclusion by 5.22. $\qquad \square$

Proposition 5.36. *Let $a \in \mathfrak{m}$ be a nonzero divisor in A. For a finitely generated A-module M for which a is a nonzero divisor we have*

$$\operatorname{Ext}_A^i(k, M) \simeq \operatorname{Ext}_{A/(a)}^{i-1}(k, M/aM).$$

Proof. Factor the functor $\operatorname{Hom}_A(k, -)$

$$\{A - mod\} \xrightarrow{\operatorname{Hom}_A(A/(a), -)} \{A/(a) - mod\} \xrightarrow{\operatorname{Hom}_{A/(a)}(k, -)} \{k - mod\}.$$

Note that $\operatorname{Hom}_A(A/(a), -)$ transforms injective A-modules into injective $A/(a)$-modules and preserves essential extensions. Consider a minimal injective resolution of M

$$0 \longrightarrow M \longrightarrow I^0 \longrightarrow I^1 \longrightarrow \cdots$$

and consider the complex of injective $A/(a)$-modules

$$J^{\boldsymbol{\cdot}} = \operatorname{Hom}_A(A/(a), I^{\boldsymbol{\cdot}}).$$

Note $J^0 = 0$ and $H^i(J^{\boldsymbol{\cdot}}) = \operatorname{Ext}_A^i(A/(a), M)$. Using the free resolution

$$0 \longrightarrow A \xrightarrow{a} A \longrightarrow A/(a) \longrightarrow 0$$

we get

$$H^i(J^{\boldsymbol{\cdot}}) = \begin{cases} M/aM & i = 1 \\ 0 & i > 1, \end{cases}$$

thus $J^{\boldsymbol{\cdot}}[1]$ is an injective resolution of the $A/(a)$-module M/aM. Whence

$$\begin{aligned}
\operatorname{Ext}_{A/(a)}^i(k, M/aM) &\simeq H^i(\operatorname{Hom}_{A/(a)}(k, J^{\boldsymbol{\cdot}}[1])) \\
&\simeq H^{i+1}(\operatorname{Hom}_{A/(a)}(k, \operatorname{Hom}_A(A/(a), I^{\boldsymbol{\cdot}}))) \\
&\simeq H^{i+1}(\operatorname{Hom}_A(k, I^{\boldsymbol{\cdot}})). \qquad \square
\end{aligned}$$

Definition 5.37. The local ring A is called a *Gorenstein* local ring if the A-module A has a finite injective resolution.

Remark 5.38. A regular local ring is Gorenstein. A is Gorenstein if and only if the completion $\hat{A}$ is Gorenstein. If A is a Gorenstein ring and $\mathfrak{p}$ a prime ideal in A then $A_\mathfrak{p}$ is a Gorenstein ring.

Corollary 5.39. *Let $a \in \mathfrak{m}$ be a nonzero divisor in the local ring A. Then $A/(a)$ is a Gorenstein local ring if and only if A is a Gorenstein ring.*

Proof. Follows from 5.35 and 5.36. $\qquad \square$

Theorem 5.40. *Let A be a Gorenstein local ring. Then A has a minimal injective resolution $A \to E^{\bullet}(A)$, with*

$$E^i(A) = \bigoplus_{\{\mathfrak{p}|\dim A_{\mathfrak{p}}=i\}} E(A/\mathfrak{p}).$$

Proof. Collect together 5.35, 5.36 and 5.38. □

Chapter 6

Local Cohomology

6.1 Basic properties

For an injective system of modules $\to M_n \to M_{n+1} \to$ the *inductive limit* $\varinjlim M_n$ is the factor module of $\bigoplus M_n$ by the submodule generated by elements $x_n - x_{n+1}$ where $x_n \mapsto x_{n+1}$.

Let A denote a noetherian ring and $\mathfrak{J}$ an ideal in A.

Definition 6.1. For an A-module M put

$$\Gamma_{\mathfrak{J}}(M) = \{x \in M \mid \mathfrak{J}^r x = 0 \text{ for some } r \in \mathbb{N}\}.$$

The functor $M \mapsto \Gamma_{\mathfrak{J}}(M)$ is left exact and its i'th derived functor $M \to H^i_{\mathfrak{J}}(M)$, $i \in \mathbb{N}$ is called the *i'th local cohomology group* with support in $\mathfrak{J}$.

It follows immediately that for an A-module M

$$H^i_{\mathfrak{J}}(M) = \varinjlim \operatorname{Ext}^i_A(A/\mathfrak{J}^r, M). \tag{6.1}$$

Local cohomology depends only on $V(\mathfrak{J})$, we have

$$H^{\cdot}_{\mathfrak{J}}(M) \simeq H^{\cdot}_{\mathfrak{J}}(M) \qquad \text{if } V(\mathfrak{J}) = V(\mathfrak{J}).$$

Let $a. = (a_1, \ldots, a_d)$ be a sequence of elements in A and

$$K.(a^r_{\cdot}) = K.(a^r_d) \otimes_A \cdots \otimes_A K.(a^r_1)$$

the Koszul complex. If $V(a_1, \ldots, a_d) = V(\mathfrak{J})$ then

$$H^i_{\mathfrak{J}}(M) = H^i(\varinjlim \operatorname{Hom}_A(K.(a.^r), M))$$

where $K.(a^{r+1}_i) \to K.(a^r_i)$ is the identity on 0-chains and multiplication with a_i on 1-chains.

Proposition 6.2. *If A is a noetherian ring, then $H_{\mathfrak{J}}^i(-)$ preserves direct sums.*

Proof. It follows from 5.10 that for a finitely generated A-module N, the functor $\mathrm{Ext}_A^i(N, -)$ preserves direct sums. $\qquad\square$

Proposition 6.3. *Suppose A is a noetherian ring. Then for an A-module M we have*

(1) $\mathrm{Supp}(H_{\mathfrak{J}}^\bullet(M)) \subseteq V(\mathfrak{J}) \cap \mathrm{Supp}(M)$.

(2) *If $\mathrm{Supp}(M) \subseteq V(\mathfrak{J})$ then $H_{\mathfrak{J}}^0(M) = M$ and $H_{\mathfrak{J}}^i(M) = 0$ for $i > 0$.*

Proof. The first part is trivial. To prove the second part, consider the full subcategory $\{mod_{\mathfrak{J}}\}$ of the category of A-modules consisting of modules with support in $V(\mathfrak{J})$. Note first that the restriction of $M \mapsto \Gamma_{\mathfrak{J}}(M)$ to $\{mod_{\mathfrak{J}}\}$ is the identity. Next let us prove that if M is in $\{mod_{\mathfrak{J}}\}$ then there exists an embedding $M \to E$ where E is an injective A-module contained in $\{mod_{\mathfrak{J}}\}$. Consider an injective envelope $M \to E$. Since any localization functor preserves injective envelopes it follows that $\mathrm{Supp}(E) = \mathrm{Supp}(M) \subseteq V(\mathfrak{J})$. The combination of these two facts and a standard argument on derived functors conclude the proof. $\qquad\square$

Proposition 6.4. *Let $\mathfrak{a}$ be an ideal in A. For any $A/\mathfrak{a}$-module M, we have*

$$H_{\mathfrak{J}+\mathfrak{a}/\mathfrak{a}}^\bullet(M) = H_{\mathfrak{J}}^\bullet(M).$$

Proof. Clearly, the composite of the inclusion $\{A/\mathfrak{a}-mod\} \to \{A-mod\}$ and $\Gamma_{\mathfrak{J}}$ is $\Gamma_{\mathfrak{J}+\mathfrak{a}/\mathfrak{a}}$. Thus it suffices to prove that if E is an injective $A/\mathfrak{a}$-module, then $H_{\mathfrak{J}}^i(E) = 0$, $i > 0$. It suffices to treat the case $E = E_{A/\mathfrak{a}}(A/\mathfrak{p})$ where $\mathfrak{p}$ is a prime ideal in $A/\mathfrak{a}$. In case $\mathfrak{J} \subseteq \mathfrak{p}$ we have $\mathrm{Supp}(E_{A/\mathfrak{a}}(A/\mathfrak{p})) \subseteq V(\mathfrak{J})$ and the result follows from 6.3. Suppose $\mathfrak{J} \not\subseteq \mathfrak{p}$ and choose $a \in \mathfrak{J} - \mathfrak{p}$. Scalar multiplication with a on

$$H_{\mathfrak{J}}^i(E_{A/\mathfrak{a}}(A/\mathfrak{p})), \qquad i \in \mathbb{N}$$

is an isomorphism. This module has support in $V(\mathfrak{J})$ and since $a \in \mathfrak{J}$ this module must be zero. $\qquad\square$

Proposition 6.5. *Suppose A is noetherian and let N be a finitely generated A-module, $\hat{N}$ its $\mathfrak{J}$-adic completion. Then*

$$H_{\mathfrak{J}}^\bullet(N) \simeq H_{\hat{\mathfrak{J}}}^\bullet(\hat{N}).$$

Proof. By (6.1) it suffices to prove that

$$\mathrm{Ext}_A^{\cdot}(A/\mathfrak{J}^r, N) \simeq \mathrm{Ext}_{\hat{A}}^{\cdot}(\hat{A}/\hat{\mathfrak{J}}^r, \hat{N}).$$

For this consider a resolution $K.$ of $A/\mathfrak{J}^r$ by finitely generated free A-modules, then

$$\mathrm{Hom}_{\hat{A}}(K.\otimes_A \hat{A}, \hat{N}) \simeq \mathrm{Hom}_A(K., N) \otimes_A \hat{A}$$

and whence

$$\begin{aligned}\mathrm{Ext}_{\hat{A}}^{\cdot}(\hat{A}/\hat{\mathfrak{J}}^r, \hat{N}) &\simeq \mathrm{Ext}_A^{\cdot}(A/\mathfrak{J}^r, N) \otimes_A \hat{A}\\ &\simeq \mathrm{Ext}_A^{\cdot}(A/\mathfrak{J}^r, N).\end{aligned} \qquad \square$$

Proposition 6.6. *Suppose A is noetherian and let M be any A-module. Given ideals $\mathfrak{I}$, $\mathfrak{J}$ there is a long exact sequence*

$$\cdots \longrightarrow H_{\mathfrak{I}\cap\mathfrak{J}}^{i-1}(M) \longrightarrow H_{\mathfrak{I}+\mathfrak{J}}^{i}(M) \longrightarrow H_{\mathfrak{I}}^{i}(M) \oplus H_{\mathfrak{J}}^{i}(M)$$
$$\longrightarrow H_{\mathfrak{I}\cap\mathfrak{J}}^{i}(M) \longrightarrow H_{\mathfrak{I}+\mathfrak{J}}^{i+1}(M) \longrightarrow \cdots.$$

Proof. Consider the exact sequence

$$0 \longrightarrow A/\mathfrak{I}^n \cap \mathfrak{J}^n \longrightarrow A/\mathfrak{I}^n \oplus A/\mathfrak{J}^n \longrightarrow A/\mathfrak{I}^n + \mathfrak{J}^n \longrightarrow 0.$$

This gives us

$$\cdots \longrightarrow \varinjlim \mathrm{Ext}_A^i(A/\mathfrak{I}^n \cap \mathfrak{J}^n, M)$$
$$\longrightarrow \varinjlim \mathrm{Ext}_A^i(A/\mathfrak{I}^n, M) \oplus \varinjlim \mathrm{Ext}_A^i(A/\mathfrak{J}^n, M)$$
$$\longrightarrow \varinjlim \mathrm{Ext}_A^i(A/\mathfrak{I}^n + \mathfrak{J}^n, M) \longrightarrow \cdots.$$

Clearly

$$\mathfrak{I}^{2n} + \mathfrak{J}^{2n} \subseteq (\mathfrak{I}+\mathfrak{J})^{2n} \subseteq \mathfrak{I}^n + \mathfrak{J}^n$$

and by the Artin–Rees lemma, 1.33,

$$(\mathfrak{I}\cap\mathfrak{J})^n \subseteq \mathfrak{I}^n \cap \mathfrak{J}^n \subseteq (\mathfrak{I}\cap\mathfrak{J})^{n-n_0}.$$

Conclusion by obvious isomorphisms. $\qquad \square$

6.2 Local cohomology and dimension

Let A denote a noetherian local ring and $\mathfrak{I}$ an ideal of A contained in the maximal ideal $\mathfrak{m}$.

Theorem 6.7. *Let $M \neq 0$ be a finitely generated A-module. Then*

$$H_{\mathfrak{I}}^i(M) = 0, \qquad i > \dim(M).$$

Proof. Induction on $d = \dim(M)$. The case $d = 0$ follows from 6.3. In general consider the exact sequence

$$0 \longrightarrow \Gamma_{\mathfrak{I}}(M) \longrightarrow M \longrightarrow M' \longrightarrow 0.$$

We have $H_{\mathfrak{I}}^i(\Gamma_{\mathfrak{I}}(M)) = 0$ for $i > 0$ as it follows from 6.3. Thus it suffices to prove that $H_{\mathfrak{I}}^i(M') = 0$ for $i > d$. Suppose $M' \neq 0$, then $\mathfrak{I} \not\subseteq \bigcup_{\mathfrak{p} \in \mathrm{Ass}(M')} \mathfrak{p}$. Thus we can choose $a \in \mathfrak{I}$ being a nonzero divisor for M'. The exact sequence

$$0 \longrightarrow M' \xrightarrow{\ a\ } M' \longrightarrow M'' \longrightarrow 0$$

gives rise to an exact sequence

$$H_{\mathfrak{I}}^{i-1}(M'') \longrightarrow H_{\mathfrak{I}}^i(M') \xrightarrow{\ a\ } H_{\mathfrak{I}}^i(M').$$

We have

$$\dim(M'') = \dim(M') - 1 \leq \dim(M) - 1 = d - 1.$$

Thus the induction hypothesis gives $H_{\mathfrak{I}}^{i-1}(M'') = 0$ for $i - 1 > d - 1$. So for $i > d$, a is a nonzero divisor for $H_{\mathfrak{I}}^i(M')$. This implies $H_{\mathfrak{I}}^i(M') = 0$, since $a \in \mathfrak{I}$ and $\mathrm{Supp}(H_{\mathfrak{I}}^i(M')) \subseteq V(\mathfrak{I})$. $\qquad\square$

6.3 Local cohomology and depth

Let A denote a noetherian local ring and $\mathfrak{I}$ an ideal of A contained in the maximal ideal $\mathfrak{m}$.

Theorem 6.8. *Let $M \neq 0$ be a finitely generated A-module. Then*

$$H_{\mathfrak{I}}^i(M) = 0, \qquad i < \mathrm{depth}_{\mathfrak{I}}(M),$$
$$H_{\mathfrak{I}}^d(M) \neq 0, \qquad d = \mathrm{depth}_{\mathfrak{I}}(M).$$

Proof. The first part follows immediately from 6.10 below. The second part is by induction on $d = \operatorname{depth}_{\mathfrak{J}} M$. The case $d = 0$ follows from 2.15. If $d > 0$ choose $a \in \mathfrak{J}$ a nonzero divisor for M. Consider the exact sequence

$$H_{\mathfrak{J}}^{d-1}(M) \longrightarrow H_{\mathfrak{J}}^{d-1}(M/aM) \longrightarrow H_{\mathfrak{J}}^{d}(M) \xrightarrow{\ a\ } H_{\mathfrak{J}}^{d}(M)$$

to draw the conclusion. $\qquad \square$

Corollary 6.9. *Let $M \neq 0$ be a finitely generated A-module. Then*

$$\operatorname{depth}_{\mathfrak{J}}(M) = \inf\{i \mid H_{\mathfrak{J}}^{i}(M) \neq 0\}.$$

Proof. Straightforward from 6.8. $\qquad \square$

Lemma 6.10. *Let $M \neq 0$ be a finitely generated A-module. Then for any finitely generated A-module N with $\operatorname{Supp}(N) \subseteq V(\mathfrak{J})$ we have*

$$\operatorname{Ext}_{A}^{i}(N, M) = 0, \qquad i < \operatorname{depth}_{\mathfrak{J}}(M).$$

Proof. Induction on $d = \operatorname{depth}_{\mathfrak{J}}(M)$. We may suppose that $d \geq 1$, so choose $a \in \mathfrak{J}$ a nonzero divisor for M. Consider the long exact sequence

$$\operatorname{Ext}_{A}^{i-1}(N, M/aM) \longrightarrow \operatorname{Ext}_{A}^{i}(N, M) \xrightarrow{\ a\ } \operatorname{Ext}_{A}^{i}(N, M)$$

to see that a is a nonzero divisor for $\operatorname{Ext}_{A}^{i}(N, M)$, $i < d$. On the other hand, $\operatorname{Supp}(\operatorname{Ext}_{A}^{i}(N, M)) \subseteq V(\mathfrak{J})$ and $a \in \mathfrak{J}$. Thus $\operatorname{Ext}_{A}^{i}(N, M) = 0$. $\qquad \square$

6.4 Support in the maximal ideal

Let A denote a noetherian local ring with maximal ideal $\mathfrak{m}$ and residue field k.

Proposition 6.11. *Let M be a finitely generated A-module. Then $H_{\mathfrak{m}}^{i}(M)$ is artinian for all $i \in \mathbb{N}$.*

Proof. Let $E^{\bullet}$ be a minimal injective resolution of M. Note first that for any prime ideal $\mathfrak{p}$ in A, we have

$$\Gamma_{\mathfrak{m}}(E(A/\mathfrak{p})) = \begin{cases} 0 & \text{for } \mathfrak{p} \neq \mathfrak{m} \\ E(A/\mathfrak{m}) & \text{for } \mathfrak{p} = \mathfrak{m}. \end{cases}$$

It follows from 5.24 that each component of the complex $\Gamma_{\mathfrak{m}}(E^{\bullet})$ is a direct sum of finitely many copies of $E(A/\mathfrak{m})$, thus by 5.18 $\Gamma_{\mathfrak{m}}(E^{\bullet})$ is a complex of artinian modules. $\qquad \square$

Proposition 6.12. *Let $M \neq 0$ be a finitely generated A-module. Then*

$$H_{\mathfrak{m}}^d(M) \neq 0, \qquad d = \dim(M).$$

Proof. Induction on d. The case $d = 0$ being trivial, suppose $d = 1$. Consider the exact sequence

$$0 \longrightarrow \Gamma_{\mathfrak{m}}(M) \longrightarrow M \longrightarrow M' \longrightarrow 0.$$

Note, $H_{\mathfrak{m}}^1(M) = H_{\mathfrak{m}}^1(M')$ and $\mathrm{depth}(M) = \mathrm{depth}(M')$. Thus we may assume $\mathrm{depth}\, M = 1$ and conclude by 6.8.

Suppose $d > 1$. We may assume that A is complete since $H_{\mathfrak{m}}^{\cdot}(M) = H_{\hat{\mathfrak{m}}}^{\cdot}(\hat{M})$ as it follows from 6.5. Let E be the injective envelope of the residue field k and for a module P, let $P^{\vee} = \mathrm{Hom}_A(P, E)$ be the Matlis dual. Proceeding as above we may assume $\mathrm{depth}(M) \neq 0$. For $a \in \mathfrak{m}$ a nonzero divisor for M we have an exact sequence

$$H_{\mathfrak{m}}^d(M)^{\vee} \longrightarrow H_{\mathfrak{m}}^{d-1}(M/aM)^{\vee} \longrightarrow H_{\mathfrak{m}}^{d-1}(M)^{\vee} \xrightarrow{\ a\ } H_{\mathfrak{m}}^{d-1}(M)^{\vee}.$$

Suppose $H_{\mathfrak{m}}^d(M) = 0$ and let us proceed to find a contradiction. By the induction hypothesis we have $H_{\mathfrak{m}}^{d-1}(M/aM) \neq 0$. Thus we have proved

$$\mathfrak{m} - \bigcup_{\mathfrak{p} \in \mathrm{Ass}(M)} \mathfrak{p} \subseteq \bigcup_{\mathfrak{q} \in \mathrm{Ass}(H_{\mathfrak{m}}^{d-1}(M)^{\vee})} \mathfrak{q}.$$

That is, $\mathfrak{m}$ is contained in the union of the prime ideals which are either associated to M or $H_{\mathfrak{m}}^{d-1}(M)^{\vee}$. Thus $\mathfrak{m}$ is associated to $H_{\mathfrak{m}}^{d-1}(M)^{\vee}$. Note that any element in $H_{\mathfrak{m}}^{d-1}(M)^{\vee}$ which is annihilated by $\mathfrak{m}$ is annihilated by a, and whence comes from $H_{\mathfrak{m}}^{d-1}(M/aM)^{\vee}$, thus we have proved that

$$\mathfrak{m} \in \mathrm{Ass}(H_{\mathfrak{m}}^{d-1}(M/aM)^{\vee})$$

contradicting the following lemma.   □

Lemma 6.13. *Let $N \neq 0$ be a finitely generated A-module of dimension $d \neq 0$. Then*

$$\mathfrak{m} \notin \mathrm{Ass}(H_{\mathfrak{m}}^d(N)^{\vee}).$$

Proof. Consider the exact sequence

$$0 \longrightarrow \Gamma_{\mathfrak{m}}(N) \longrightarrow N \longrightarrow N' \longrightarrow 0.$$

We have $H_{\mathfrak{m}}^n(N) = H_{\mathfrak{m}}^n(N')$, $\dim N = \dim N'$ and $\dim N' \neq 0$. For any $a \in \mathfrak{m}$ nonzero divisor for N we have an exact sequence

$$H_{\mathfrak{m}}^d(N/aN)^\vee \longrightarrow H_{\mathfrak{m}}^d(N)^\vee \overset{a}{\longrightarrow} H_{\mathfrak{m}}^d(N)^\vee.$$

It follows from 6.7 that a is a nonzero divisor for $H_{\mathfrak{m}}^d(N)^\vee$, thus $\mathfrak{m}$ cannot be associated to $H_{\mathfrak{m}}^d(N)^\vee$. $\qquad\square$

6.5   Local duality for Gorenstein rings

Let A denote a noetherian local ring with maximal ideal $\mathfrak{m}$ and residue field k. Let E be the injective envelope of k. We shall assume that A is Gorenstein, i.e., the module A has a finite injective resolution. Let $d = \dim(A)$, then we have

$$H_{\mathfrak{m}}^i(A) = \begin{cases} 0, & i \neq d \\ E, & i = d \end{cases}$$

as it follows immediately by considering a minimal injective resolution of A, 5.40.

For an A-module M and $i \in \mathbb{N}$ we have the Yoneda pairing

$$H_{\mathfrak{m}}^{d-i}(M) \times \mathrm{Ext}_A^i(M, A) \to H_{\mathfrak{m}}^d(A).$$

Identifying $H_{\mathfrak{m}}^d(A)$ with E we deduce a map ($-^\vee$ denotes the Matlis dual)

$$H_{\mathfrak{m}}^{d-i}(M) \to \mathrm{Ext}_A^i(M, A)^\vee.$$

Theorem 6.14. *For a Gorenstein local ring A of dimension d, we have a natural isomorphism*

$$H_{\mathfrak{m}}^{d-i}(M) \to \mathrm{Ext}_A^i(M, A)^\vee \tag{6.2}$$

for any finitely generated A-module M and any $i \in \mathbb{N}$.

Proof. Note that (6.2) is an isomorphism for trivial reasons when $i = d$ and $M = A$. Note that for $i = d$ both functors in (6.2) are right exact. Considering for M a finite presentation of finitely generated free modules, it follows that (6.2) is an isomorphism in case $i = 0$. Note that both functors in (6.2) are zero for $M = A$, $i \neq 0$. It is now straightforward to prove

by induction on i, that (6.2) is an isomorphism by using the short exact
sequence of the form

$$0 \longrightarrow N \longrightarrow A^r \longrightarrow M \longrightarrow 0$$

and the commutative ladder formed by the long exact sequences arising in
the two functors and the transformations relating them.   □

Chapter 7

Dualizing Complexes

7.1 Complexes of injective modules

By a *complex* we shall understand a (cochain)-complex of modules over a fixed ring A, $\partial^n : X^n \to X^{n+1}$, $n \in \mathbb{Z}$. A complex $X^{\cdot}$ is *bounded below* if $X^n = 0$ for $n \ll 0$ and *bounded above* if $X^n = 0$ for $0 \ll n$. A complex is *bounded* if it is both bounded below and above. Recall that if $X^{\cdot}$ and $E^{\cdot}$ are complexes we define the *Hom complex* $\mathrm{Hom}_A^{\cdot}(X^{\cdot}, E^{\cdot})$ by

$$\mathrm{Hom}_A^n(X^{\cdot}, E^{\cdot}) = \prod_{i \in \mathbb{Z}} \mathrm{Hom}_A(X^i, E^{i+n}),$$

$$\partial^n(f)^i = \partial_E^{i+n} f^i - (-1)^n f^{i+1} \partial_X^i.$$

For $n \in \mathbb{Z}$, $E^{\cdot}[n]$ denotes the *decalage complex* whose i'th cochain modules is E^{i+n} and whose i'th differential is $(-1)^n \partial^{i+n}$. With this notation we can interpret $H^n(\mathrm{Hom}_A^{\cdot}(X^{\cdot}, E^{\cdot}))$ as homotopy classes of morphisms from $X^{\cdot}$ to $E^{\cdot}[n]$.

If convenient we also use (chain)-complexes $X_{\cdot}$, $\partial_n : X_n \to X_{n-1}$ with the convention $X^n = X_{-n}$, $\partial^n = \partial_{-n}$.

Lemma 7.1. *Let $E^{\cdot}$ be a bounded below complex of injective modules. If $X^{\cdot}$ is a complex with $H^{\cdot}(X^{\cdot}) = 0$, then any morphism $f^{\cdot} : X^{\cdot} \to E^{\cdot}$ is homotopic to zero.*

Proof. We shall proceed by induction. Suppose $s^i : X^i \to E^{i-1}$ is constructed for $i \le n$, such that

$$s^i \partial^{i-1} + \partial^{i-2} s^{i-1} = f^{i-1}.$$

Consider $f^n - \partial^{n-1} s^n$ and note that $(f^n - \partial^{n-1} s^n)\partial^{n-1} = 0$. It follows that $f^n - \partial^{n-1} s^n$ can be factored through $X^n \to \mathrm{Im}(\partial^n)$. Using that E^n is injective we can find $s^{n+1} : X^{n+1} \to E^n$ such that

$$f^n - \partial^{n-1} s^n = s^{n+1} \partial^n. \qquad \square$$

Definition 7.2. A morphism of complexes $f^\bullet : X^\bullet \to Y^\bullet$ is called a *quasi-isomorphism* if $H^i(f^\bullet) : H^i(X^\bullet) \to H^i(Y^\bullet)$ is an isomorphism for all $i \in \mathbb{Z}$.

Proposition 7.3. *Let $E^\bullet$ be a bounded below complex of injective modules. Then for any quasi-isomorphism $f^\bullet : X^\bullet \to Y^\bullet$*

$$\operatorname{Hom}_A^\bullet(f^\bullet, 1) : \operatorname{Hom}_A^\bullet(Y^\bullet, E^\bullet) \to \operatorname{Hom}_A^\bullet(X^\bullet, E^\bullet)$$

is a quasi-isomorphism.

Proof. Let us recall the *mapping cone* construction, $C^\bullet(f^\bullet)$ denotes a complex given by

$$C^n(f^\bullet) = X^{n+1} \oplus Y^n$$
$$\partial^n(x, y) = (-\partial^{n+1}(x), \partial^n(y) + f^{n+1}(x)).$$

Note that we have a canonical exact sequence

$$0 \longrightarrow Y^\bullet \longrightarrow C^\bullet(f^\bullet) \longrightarrow X^\bullet[1] \longrightarrow 0,$$

which gives the long exact sequence on cohomology

$$H^{i-1}(Y^\bullet) \longrightarrow H^{i-1}(C^\bullet(f^\bullet)) \longrightarrow H^i(X^\bullet)$$
$$\xrightarrow{H^i(f^\bullet)} H^i(Y^\bullet) \longrightarrow H^i(C^\bullet(f^\bullet)),$$

thus $f^\bullet$ is a quasi-isomorphism if and only if $H^\bullet(C^\bullet(f^\bullet)) = 0$. To conclude the proof, apply the functor $\operatorname{Hom}_A^\bullet(-, E^\bullet)$ to the short exact sequence above, form the resulting long exact sequence on cohomology and apply 7.1. $\square$

The proposition above may be reformulated as follows.

Corollary 7.4. *Let $f^\bullet : X^\bullet \to Y^\bullet$ be a quasi-isomorphism and $g^\bullet : X^\bullet \to E^\bullet$ a morphism into a bounded below complex of injective modules. Then there exists $h^\bullet : Y^\bullet \to E^\bullet$ such that $g^\bullet$ is homotopic to the composition $h^\bullet \circ f^\bullet$. Moreover $h^\bullet$ is unique up to homotopy.*

Proof. The quasi-isomorphism $f^{\cdot}$ induces a quasi-isomorphism

$$\mathrm{Hom}_A^{\cdot}(Y^{\cdot}, E^{\cdot}) \to \mathrm{Hom}_A^{\cdot}(X^{\cdot}, E^{\cdot}).$$

In particular

$$H^0(\mathrm{Hom}_A^{\cdot}(Y^{\cdot}, E^{\cdot})) \to H^0(\mathrm{Hom}_A^{\cdot}(X^{\cdot}, E^{\cdot}))$$

is an isomorphism. □

Corollary 7.5. *Any quasi-isomorphism* $f^{\cdot} : D^{\cdot} \to E^{\cdot}$ *of bounded below complexes of injective modules is a homotopy equivalence, i.e., there exists a morphism* $h^{\cdot} : E^{\cdot} \to D^{\cdot}$ *such that both composites are homotopic to the identities.*

Proof. According to 7.4 we can find $h^{\cdot} : E^{\cdot} \to D^{\cdot}$ such that $h^{\cdot} \circ f^{\cdot}$ is homotopic to $1_{D^{\cdot}}$. Since $h^{\cdot}$ is a quasi-isomorphism, we may find $g^{\cdot} : D^{\cdot} \to E^{\cdot}$ such that $g^{\cdot} \circ h^{\cdot}$ is homotopic to $1_{E^{\cdot}}$. It follows that $g^{\cdot}$ is homotopic to $f^{\cdot}$ and finally that $f^{\cdot} \circ h^{\cdot}$ is homotopic to $1_{E^{\cdot}}$. □

Proposition 7.6. *For any bounded below complex $Z^{\cdot}$ there exists a bounded below complex of injective modules $E^{\cdot}$ and a quasi-isomorphism*

$$f^{\cdot} : Z^{\cdot} \to E^{\cdot}.$$

Proof. Suppose $f^{\cdot}$ has already been constructed up to level n. That is, we construct the diagram

$$
\begin{array}{ccccccccc}
Z^{n-1} & \longrightarrow & Z^n & \longrightarrow & \mathrm{Cok}(\partial_Z^{n-1}) & \overset{\epsilon}{\longrightarrow} & \mathrm{Ker}(\partial_Z^{n+1}) & \longrightarrow & Z^{n+1} \\
\downarrow{\scriptstyle f^{n-1}} & & \downarrow{\scriptstyle f^n} & & \downarrow{\scriptstyle f^{n'}} & & \downarrow & & \downarrow{\scriptstyle f^{n+1}} \\
E^{n-1} & \longrightarrow & E^n & \longrightarrow & \mathrm{Cok}(\partial_E^{n-1}) & \longrightarrow & Y^n & \longrightarrow & E^{n+1}
\end{array}
$$

with the property that $H^i(Z^{\cdot}) \to H^i(E^{\cdot})$ is an isomorphism for $i < n$. Suppose furthermore this is done such that the induced map $H^n(Z^{\cdot}) \to \mathrm{Cok}(\partial_E^{n-1})$ is an injection. Insert Y^n and the two arrows with target Y^n such that the resulting square is a pushout, i.e., choose $Y^n = \mathrm{Cok}((f^{n'}, -\epsilon))$. Next, imbed Y^n into an injective module E^{n+1} and insert the arrow f^{n+1}. The construction gives rise to the exact sequence

$$0 \longrightarrow H^n(Z^{\cdot}) \longrightarrow \mathrm{Cok}(\partial_E^{n-1}) \longrightarrow Y^n \longrightarrow H^{n+1}(Z^{\cdot}) \longrightarrow 0.$$

We leave it to the reader to check that $H^n(Z^{\cdot}) \to H^n(E^{\cdot})$ is an isomorphism and $H^{n+1}(Z^{\cdot}) \to \mathrm{Cok}(\partial_E^n)$ is an injection. □

Remark 7.7. Let $f^\cdot : X^\cdot \to Y^\cdot$ be an arbitrary morphism of complexes. Then $f^\cdot$ is a quasi-isomorphism if and only if for all $n \in \mathbb{Z}$ the following diagram is both a pullback and a pushout

$$
\begin{array}{ccc}
\mathrm{Cok}(\partial_X^{n-1}) & \xrightarrow{\epsilon_X} & \mathrm{Ker}(\partial_X^{n+1}) \\
\downarrow{f^{n\prime}} & & \downarrow{f^{n+1\prime\prime}} \\
\mathrm{Cok}(\partial_Y^{n-1}) & \xrightarrow{\epsilon_Y} & \mathrm{Ker}(\partial_Y^{n+1})
\end{array}
$$

that is, $\mathrm{Cok}(\partial_X^{n-1}) = \mathrm{Ker}((f^{n+1\prime\prime}, -\epsilon_Y))$ and $\mathrm{Ker}(\partial_Y^{n+1}) = \mathrm{Cok}((f^{n\prime}, -\epsilon_X))$.

Definition 7.8. A complex $E^\cdot$ of injective modules is called *minimal* if $E^\cdot$ is bounded below and for all $n \in \mathbb{Z}$, $\mathrm{Ker}(\partial^n) \to E^n$ is an injective envelope.

Proposition 7.9. *Any bounded below complex $Z^\cdot$ admits a quasi-isomorphism into a minimal injective complex.*

Proof. Consider the construction made in the proof of 7.6. With the same notation, the module Y^n is necessarily equal to $\mathrm{Ker}(\partial_E^{n+1})$ as it follows from 7.7. Thus it suffices to choose $Y^n \to E^{n+1}$ to be an injective envelope of Y^n. $\qquad\square$

Proposition 7.10. *Let $f^\cdot : D^\cdot \to E^\cdot$ be a quasi-isomorphism between minimal injective complexes. Then $f^\cdot$ is an isomorphism of complexes.*

Proof. By induction on i we get

$$\mathrm{Ker}(\partial_D^i) \simeq \mathrm{Ker}(\partial_E^i)$$

giving that f^i is an isomorphism of extensions. $\qquad\square$

Remark 7.11. Let $E^\cdot$ be a minimal injective complex and let $\mathfrak{m}$ be any maximal ideal in A. Using that $A/\mathfrak{m}$ is a simple module we get that the complex $\mathrm{Hom}_A^\cdot(A/\mathfrak{m}, E^\cdot)$ has zero differentials.

More precisely, a bounded below complex $E^\cdot$ of injective modules is minimal if and only if the complex $\mathrm{Hom}_{A_\mathfrak{p}}^\cdot(k_\mathfrak{p}, E_\mathfrak{p}^\cdot)$ has zero differentials for all $\mathfrak{p} \in \mathrm{Spec}(A)$.

Remark 7.12. The propositions 7.3–7.9 have the following dual forms which we state for convenience. Let $P^\cdot$ be a bounded above complex of projective modules. Then for any quasi-isomorphism $X^\cdot \to Y^\cdot$ there is induced a quasi-isomorphism $\mathrm{Hom}_A^{\bullet}(P^{\bullet}, X^{\bullet}) \to \mathrm{Hom}_A^{\bullet}(P^{\bullet}, Y^{\bullet})$. Any quasi-isomorphism between bounded above complexes of projective modules is a homotopy equivalence. For a bounded above complex $X^{\bullet}$ there exists a quasi-isomorphism $P^{\bullet} \to X^{\bullet}$ where $P^{\bullet}$ is a bounded above complex of projective modules. If the ring A is a noetherian local ring with maximal ideal $\mathfrak{m}$, and $X^{\bullet}$ is bounded above with finitely generated cochains, then the complex $P^{\bullet}$ may be chosen minimal, i.e., P^i is a finitely generated free module and $\partial^i(P^i) \subseteq \mathfrak{m}P^{i+1}$.

7.2 Complexes with finitely generated cohomology

Throughout this section A denotes a noetherian ring. By a module is understood an A-module and by a complex is understood a complex of A-modules. Given a complex $E^{\bullet}$ of injective modules and a module M then for all $i \in \mathbb{Z}$ we put

$$\mathrm{Ext}_A^i(M, E^{\bullet}) = H^i(\mathrm{Hom}_A^{\bullet}(M, E^{\bullet})).$$

Proposition 7.13. *Let $X^{\bullet}$ be a bounded above complex and $E^{\bullet}$ a bounded below complex of injective modules, and suppose both complexes have finitely generated cohomology modules. Then $\mathrm{Hom}_A^{\bullet}(X^{\bullet}, E^{\bullet})$ has finitely generated cohomology modules.*

Proof. We shall first prove that for any finitely generated module M and $p \in \mathbb{Z}$, $H^p(\mathrm{Hom}_A^{\bullet}(M, E^{\bullet}))$ are finitely generated modules. This is done by increasing induction on p. The result is clear for $p \ll 0$. For the inductive step consider the exact sequence

$$0 \longrightarrow N \longrightarrow A^m \longrightarrow M \longrightarrow 0$$

and the resulting long exact sequence

$$H^{p-1}(\mathrm{Hom}_A^{\bullet}(N, E^{\bullet})) \longrightarrow H^p(\mathrm{Hom}_A^{\bullet}(M, E^{\bullet})) \longrightarrow H^p(\mathrm{Hom}_A^{\bullet}(A^m, E^{\bullet}))$$

from which the result follows.

In the general case for fixed $p \in \mathbb{Z}$ we shall prove that for any complex $X^{\bullet}$ (with finitely generated cohomology) of the form

$$\cdots \longrightarrow X^{n-1} \xrightarrow{\ \partial^{n-1}\ } X^n \xrightarrow{\ \partial^n\ } X^{n+1} \longrightarrow 0 \longrightarrow \cdots$$

$H^p(\operatorname{Hom}^{\bullet}_A(X^{\bullet}, E^{\bullet}))$ is finitely generated. This is done by increasing induction on n. The result is clear in case $n \ll 0$. Consider the following short exact sequence of complexes

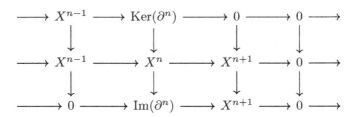

and finally the exact short sequence of complexes

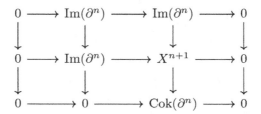

together with 7.3 to conclude the proof.   □

Proposition 7.14. *Let $X^{\bullet}$ be a bounded above complex and $E^{\bullet}$ a bounded below complex of injective modules, and suppose both complexes have finitely generated cohomology modules. Let S be a multiplicative closed subset of A then the canonical map*

$$S^{-1}\operatorname{Hom}^{\bullet}_A(X^{\bullet}, E^{\bullet}) \to \operatorname{Hom}^{\bullet}_{S^{-1}A}(S^{-1}X^{\bullet}, S^{-1}E^{\bullet})$$

is a quasi-isomorphism.

Proof. For a module M and $p \in \mathbb{Z}$ put

$$V^p(M) = H^p(S^{-1}\operatorname{Hom}^{\bullet}_A(M, E^{\bullet}))$$

and

$$U^p(M) = H^p(\operatorname{Hom}^{\bullet}_{S^{-1}A}(S^{-1}M, S^{-1}E^{\bullet}))$$

and let $\theta^p(M) : V^p(M) \to U^p(M)$ denote the natural transformation. We shall prove by increasing induction on p, that $\theta^p(M)$ is an isomorphism for all finitely generated modules M. Note that $\theta^p(A)$ is always an isomorphism. In general choose a short exact sequence

$$0 \longrightarrow N \longrightarrow A^m \longrightarrow M \longrightarrow 0$$

and consider the commutative exact diagram:

$$
\begin{array}{ccccccccc}
V^{p-1}(A^n) & \longrightarrow & V^{p-1}(N) & \longrightarrow & V^p(M) & \longrightarrow & V^p(A^n) & \longrightarrow & V^p(N) \\
\downarrow & & \downarrow & & \downarrow & & \downarrow & & \\
U^{p-1}(A^n) & \longrightarrow & U^{p-1}(N) & \longrightarrow & U^p(M) & \longrightarrow & U^p(A^n) & \longrightarrow & U^p(N) \\
\downarrow & & & & & & & & \\
0 & & & & & & & &
\end{array}
$$

This gives rise to the short exact sequence

$$\mathrm{Ker}(\theta^{p-1}(N)) \longrightarrow \mathrm{Ker}(\theta^p(M)) \longrightarrow \mathrm{Ker}(\theta^p(A^n)),$$

from which we conclude that $\theta^p(M)$ is injective for all finitely generated modules M. In particular for N as above. Thus we get a short exact sequence

$$\mathrm{Cok}(\theta^{p-1}(N)) \longrightarrow \mathrm{Cok}(\theta^p(M)) \longrightarrow \mathrm{Cok}(\theta^p(A^n)),$$

from which we conclude that $\theta^p(M)$ is surjective. For the general case, proceed as in the end of the proof of 7.13. □

Proposition 7.15. *Let $E^{\cdot}$ be a bounded below complex of injective modules with finitely generated cohomology modules. If $r, s \in \mathbb{Z}$ are such that for any prime ideal $\mathfrak{p}$ in A with residue field $k_{\mathfrak{p}} = A_{\mathfrak{p}}/\mathfrak{p}A_{\mathfrak{p}}$*

$$\mathrm{Ext}^i_{A_{\mathfrak{p}}}(k_{\mathfrak{p}}, E^{\cdot}_{\mathfrak{p}}) = 0 \qquad \text{for } i \notin [r, s],$$

then $E^{\cdot}$ is homotopy equivalent to a complex $F^{\cdot}$ of injective modules with

$$F^i = 0 \qquad \text{for } i \notin [r, s].$$

Proof. We may assume by 7.9 that $E^{\cdot}$ is a minimal injective complex. For a prime ideal $\mathfrak{p}$, $E^{\cdot}_{\mathfrak{p}}$ is still a minimal injective complex and the complex

$$\mathrm{Hom}^{\cdot}_{A_{\mathfrak{p}}}(k_{\mathfrak{p}}, E^{\cdot}_{\mathfrak{p}})$$

has zero differentials, compare 5.5 and 7.11. □

Proposition 7.16. *Let $E^{\cdot}$ be a bounded below complex of injective modules having finitely generated cohomology. For prime ideals $\mathfrak{p} \subset \mathfrak{q}$ with no prime ideal lying properly between, we have for all $i \in \mathbb{Z}$*

$$\operatorname{Ext}^i_{A_{\mathfrak{p}}}(k_{\mathfrak{p}}, E^{\cdot}_{\mathfrak{p}}) \neq 0 \quad \Rightarrow \quad \operatorname{Ext}^{i+1}_{A_{\mathfrak{q}}}(k_{\mathfrak{q}}, E^{\cdot}_{\mathfrak{q}}) \neq 0.$$

Proof. We may assume A local with maximal ideal $\mathfrak{q}$ and residue field k. Assume $\operatorname{Ext}^{i+1}_A(k, E^{\cdot}) = 0$. This implies that for any module N of finite length, $\operatorname{Ext}^{i+1}_A(N, E^{\cdot}) = 0$ as one sees by induction on length of N. Choose $a \in \mathfrak{q} - \mathfrak{p}$. The short exact sequence

$$0 \longrightarrow A/\mathfrak{p} \overset{a}{\longrightarrow} A/\mathfrak{p} \longrightarrow N \longrightarrow 0$$

gives rise to a long exact sequence

$$\operatorname{Ext}^i_A(A/\mathfrak{p}, E^{\cdot}) \overset{a}{\longrightarrow} \operatorname{Ext}^i_A(A/\mathfrak{p}, E^{\cdot}) \longrightarrow \operatorname{Ext}^{i+1}_A(N, E^{\cdot}).$$

Now it follows from Nakayama's lemma that $\operatorname{Ext}^i_A(A/\mathfrak{p}, E^{\cdot}) = 0$ and whence $\operatorname{Ext}^i_{A_{\mathfrak{p}}}(k_{\mathfrak{p}}, E^{\cdot}_{\mathfrak{p}}) = 0$. $\qquad \square$

Proposition 7.17. *Let A be a noetherian local ring with residue field k. Any bounded below complex $E^{\cdot}$ of injective modules with finitely generated cohomology for which*

$$\operatorname{Ext}^i_A(k, E^{\cdot}) = 0, \qquad for\ i > r$$

is homotopy equivalent to a bounded complex $F^{\cdot}$ of injective modules with

$$F^i = 0, \qquad for\ i > r.$$

Proof. Combine 7.15 and 7.16. $\qquad \square$

7.3   The evaluation map

Let $X^{\cdot}$ and $E^{\cdot}$ be complexes over a fixed ring A. For $n \in \mathbb{Z}$ consider the map

$$X^n \to \operatorname{Hom}^n_A(\operatorname{Hom}^{\cdot}_A(X^{\cdot}, E^{\cdot}), E^{\cdot})$$

which to the element $x_n \in X^n$ assigns the product over $i \in \mathbb{Z}$ of the maps

$$\operatorname{Hom}^i_A(X^{\cdot}, E^{\cdot}) \to E^{i+n}, \qquad f \mapsto (-1)^{in} f^n(x_n).$$

We leave it to the reader to establish that this defines a map of complexes

$$ev : X^{\cdot} \to \operatorname{Hom}_A^{\cdot}(\operatorname{Hom}_A^{\cdot}(X^{\cdot}, E^{\cdot}), E^{\cdot})$$

which we call the *evaluation* map. In particular if $X^{\cdot}$ is the complex $X^n = 0$, $n \neq 0$ and $X^0 = A$ we get the map

$$ev : A \to \operatorname{Hom}_A^{\cdot}(E^{\cdot}, E^{\cdot}).$$

In the rest of this section we shall assume that A is a noetherian ring.

Definition 7.18. A *dualizing complex* for the ring A is a bounded complex $D^{\cdot}$ of injective modules with finitely generated cohomology such that

$$ev : X^{\cdot} \to \operatorname{Hom}_A^{\cdot}(\operatorname{Hom}_A^{\cdot}(X^{\cdot}, D^{\cdot}), D^{\cdot})$$

is a quasi-isomorphism for any complex $X^{\cdot}$ with finitely generated cohomology.

Proposition 7.19. *Let $D^{\cdot}$ be a bounded complex of injective modules having finitely generated cohomology. If the canonical map*

$$ev : A \to \operatorname{Hom}_A^{\cdot}(D^{\cdot}, D^{\cdot})$$

is a quasi-isomorphism, then $D^{\cdot}$ is a dualizing complex for A.

Proof. Use the methods developed in the proof of 7.14 and 7.13 for a bounded above complex. For a general complex $X^{\cdot}$ and a fixed $p \in \mathbb{Z}$ notice that

$$\operatorname{Hom}_A^p(\operatorname{Hom}_A^{\cdot}(X^{\cdot}, D^{\cdot}), D^{\cdot})$$

only depends on the truncated above complex

$$\cdots \longrightarrow X^n \longrightarrow \operatorname{Ker}(\partial^{n+1}) \longrightarrow 0$$

for $n \gg 0$. $\qquad\square$

Corollary 7.20. *If the module A has a finite injective resolution $E^{\cdot}$, then $E^{\cdot}$ is a dualizing complex for A.*

Proof. A quasi-isomorphism $\theta : A \to E^{\cdot}$ gives rise to a quasi-isomorphism $\operatorname{Hom}_A^{\cdot}(E^{\cdot}, E^{\cdot}) \to \operatorname{Hom}_A^{\cdot}(A, E^{\cdot})$ making the following diagram commutative:

$$
\begin{array}{ccc}
A & \xrightarrow{\ ev\ } & \operatorname{Hom}_A^{\cdot}(E^{\cdot}, E^{\cdot}) \\
\downarrow{\scriptstyle \theta} & & \downarrow \\
E^{\cdot} & \xrightarrow{\ \cong\ } & \operatorname{Hom}_A^{\cdot}(A, E^{\cdot})
\end{array}
$$

$\qquad\square$

Example 7.21. The complex

$$\mathbb{Q} \to \mathbb{Q}/\mathbb{Z}$$

is a dualizing complex for $\mathbb{Z}$.

Example 7.22. Let A denote a Gorenstein local ring. Then a minimal injective resolution $E^\cdot$ of the module A is a dualizing complex for A.

Example 7.23. Let k denote a field and $E^\cdot$ a bounded complex of k-vector spaces. Then $E^\cdot$ is a dualizing complex, if and only if there exists a $d \in \mathbb{Z}$ such that

$$H^i(E^\cdot) = \begin{cases} 0, & i \neq d, \\ k, & i = d. \end{cases}$$

To see this we may assume that $E^\cdot$ consists of finitely generated vector spaces with zero differentials. We have

$$H^0(\mathrm{Hom}_A^\cdot(E^\cdot, E^\cdot)) = \bigoplus_i \mathrm{End}(E^i)$$

from which the conclusion easily follows.

Remark 7.24. Let $D^\cdot$ be a dualizing complex for A. Then any bounded complex of injective modules which is homotopy equivalent to $D^\cdot$ is a dualizing complex. For any $n \in \mathbb{Z}$, $D^\cdot[n]$ is a dualizing complex for A. For any projective module L of rank one $D^\cdot \otimes_A L$ is a dualizing complex for A.

We shall prove in section 7.8 that if $\mathrm{Spec}(A)$ is connected, then every other dualizing complex is of the form $D^\cdot[n] \otimes_A L$.

7.4   Existence of dualizing complexes

It should at once be pointed out that not all noetherian local rings have dualizing complexes. The principal tools for constructing them are exposed in this section.

Proposition 7.25. *Let A be a noetherian ring and $D^\cdot$ a dualizing complex for A. For any ideal $\mathfrak{I}$ in A,*

$$\mathrm{Hom}_A^\cdot(A/\mathfrak{I}, D^\cdot)$$

is a dualizing complex for $A/\mathfrak{I}$.

Proof. It follows immediately from 7.13 that $\mathrm{Hom}_A^{\bullet}(A/\mathfrak{I}, D^{\bullet})$ has finitely generated cohomology. Note, that if N is any $A/\mathfrak{I}$-module and E any A-module then we have a canonical isomorphism

$$\mathrm{Hom}_{A/\mathfrak{I}}(N, \mathrm{Hom}_A(A/\mathfrak{I}, E)) \to \mathrm{Hom}_A(N, E).$$

In particular if E is an injective A module then $\mathrm{Hom}_A(A/\mathfrak{I}, E)$ is an injective $A/\mathfrak{I}$-module. The rest now follows by applying the isomorphism above twice. $\square$

Corollary 7.26. *Let A be a noetherian local ring which is a quotient ring of some Gorenstein local ring. Then A has a dualizing complex.*

Proof. Combine 7.22 and 7.25 to get the result. $\square$

Theorem 7.27. *Any complete noetherian local ring has a dualizing complex.*

Proof. The theorem of Cohen, 4.17, asserts that any complete local ring is a quotient of a complete regular local ring. Thus conclusion follows from 7.26. $\square$

Proposition 7.28. *Let S be a multiplicative closed subset of a noetherian ring A. If $D^{\bullet}$ is a dualizing complex for A, then $S^{-1}D^{\bullet}$ is a dualizing complex for $S^{-1}A$.*

Proof. It follows from 5.13 that $S^{-1}D^{\bullet}$ is a complex of injective $S^{-1}A$-modules. It is now easy to conclude by applying 7.14 twice. $\square$

Proposition 7.29. *Let $A \to B$ be a morphism of noetherian rings. Suppose B is finitely generated as A-module. If $D^{\bullet}$ is a dualizing complex for A, then*

$$\mathrm{Hom}_A^{\bullet}(B, D^{\bullet})$$

is a dualizing complex for B.

Proof. For any complex $X^{\bullet}$ of B-modules we have a standard isomorphism

$$\mathrm{Hom}_B^{\bullet}(X^{\bullet}, \mathrm{Hom}_A^{\bullet}(B, D^{\bullet})) \simeq \mathrm{Hom}_A^{\bullet}(X^{\bullet}, D^{\bullet}).$$

From this follows that $\mathrm{Hom}_A^{\bullet}(B, D^{\bullet})$ is a complex of injective modules. Applying the above isomorphism twice it follows that $\mathrm{Hom}_A^{\bullet}(B, D^{\bullet})$ is a dualizing complex. $\square$

7.5   The codimension function

Throughout this section A denotes a noetherian ring.

Proposition 7.30. *Let $D^{\cdot}$ be a dualizing complex for A. For each prime ideal $\mathfrak{p}$ in A, there exists $r(\mathfrak{p}) \in \mathbb{Z}$ such that*

$$\mathrm{Ext}^i_{A_\mathfrak{p}}(k_\mathfrak{p}, D^{\cdot}_\mathfrak{p}) \simeq \begin{cases} 0, & i \neq r(\mathfrak{p}) \\ k_\mathfrak{p}, & i = r(\mathfrak{p}). \end{cases}$$

Proof. It follows from 7.28 that $D^{\cdot}_\mathfrak{p}$ is a dualizing complex for $A_\mathfrak{p}$ and whence it follows from 7.25 that $\mathrm{Hom}^{\cdot}_{A_\mathfrak{p}}(k_\mathfrak{p}, D^{\cdot}_\mathfrak{p})$ is a dualizing complex for $k_\mathfrak{p}$. Conclusion by 7.23.   □

Definition 7.31. Let $D^{\cdot}$ be a dualizing complex for A. The function

$$r : \mathrm{Spec}(A) \to \mathbb{Z}$$

defined in 7.30 is called the *codimension* function for $D^{\cdot}$.

Proposition 7.32. *Let $D^{\cdot}$ be a dualizing complex for A and r the codimension function for $D^{\cdot}$. Then for any pair of prime ideals $\mathfrak{p} \subset \mathfrak{q}$ with no prime ideal lying properly between,*

$$r(\mathfrak{q}) = r(\mathfrak{p}) + 1.$$

Proof. Follows from 7.16.   □

Corollary 7.33. *Suppose A has a dualizing complex. Then any saturated chain of prime ideals between pair of prime ideals $\mathfrak{p} \subset \mathfrak{q}$ has the same length. That is, the ring A is catenarian.*

Proof. Follows from 7.32.   □

Proposition 7.34. *Suppose $r : \mathrm{Spec}(A) \to \mathbb{Z}$ is the codimension function for some dualizing complex. Then A has a dualizing complex $D^{\cdot}$ with*

$$D^i = \bigoplus_{\mathfrak{p}, r(\mathfrak{p})=i} E(A/\mathfrak{p}).$$

Proof. Consider a minimal injective complex which is dualizing and compare with 7.9.   □

Proposition 7.35. *Let A be a noetherian local ring possessing a dualizing complex. Put $\dim(A) = d$, then A has a (special normalized) dualizing complex $D^{\cdot}$ with*

$$D^i = \bigoplus_{\substack{\mathfrak{p} \\ \dim(A/\mathfrak{p})=d-i}} E(A/\mathfrak{p}).$$

Proof. Consider a minimal injective complex which is dualizing shifted such that $\mathfrak{p} \mapsto \dim(A_{\mathfrak{p}})$ is a codimension function and compare with 7.34. $\square$

The implication in Theorem 7.30 admits as a converse implication the following important criterion being a dualizing complex.

Proposition 7.36. *Suppose A is local and $D^{\cdot}$ is a bounded complex of injective modules with finitely generated cohomology. If there exists $d \in \mathbb{Z}$ such that*

$$\mathrm{Ext}^i_A(k, D^{\cdot}) \simeq \begin{cases} 0, & i \neq d \\ k, & i = d \end{cases}$$

then $D^{\cdot}$ is a dualizing complex for A.

Proof. By 7.19 it suffices to prove that for any finitely generated A-module M the morphism

$$ev : M \to \mathrm{Hom}^{\cdot}_A(\mathrm{Hom}^{\cdot}_A(M, D^{\cdot}), D^{\cdot})$$

is a quasi-isomorphism.

For $i \in \mathbb{Z}$ put $T^i(M) = H^i(\mathrm{Hom}^{\cdot}_A(\mathrm{Hom}^{\cdot}_A(M, D^{\cdot}), D^{\cdot}))$ and $S^i(M) = 0$, $i \neq 0$, $S^0(M) = M$. Let $\theta^i(M) : S^i(M) \to T^i(M)$ denote the map induced by the evaluation map. Suppose M has finite length, then it follows by induction on the length of M and the 5-lemma that $\theta^i(M)$ is an isomorphism. We shall proceed by induction on $\dim(M)$. Consider the short exact sequence

$$0 \longrightarrow M' \longrightarrow M \longrightarrow M'' \longrightarrow 0,$$

where M' is the largest submodule of M of finite length. By the 5-lemma it suffices to prove that $\theta^i(M'')$ is an isomorphism for all $i \in \mathbb{Z}$. Note $\dim(M'') = \dim(M)$ and $\mathfrak{m} \notin \mathrm{Ass}(M'')$. Thus we may assume $\mathfrak{m} \notin \mathrm{Ass}(M)$. Choose $a \in \mathfrak{m}$ a nonzero divisor for M. The exact sequence

$$0 \longrightarrow M \overset{a}{\longrightarrow} M \longrightarrow M/aM \longrightarrow 0$$

gives rise by induction to an exact commutative diagram:

$$
\begin{array}{ccccccc}
S^{i-1}(M/aM) & \longrightarrow & S^i(M) & \xrightarrow{a} & S^i(M) & \longrightarrow & S^i(M/aM) \\
\downarrow{\simeq} & & \downarrow & & \downarrow & & \downarrow{\simeq} \\
T^{i-1}(M/aM) & \longrightarrow & T^i(M) & \xrightarrow{a} & T^i(M) & \longrightarrow & T^i(M/aM)
\end{array}
$$

Deduce first an exact sequence

$$
\operatorname{Ker}(\theta^i(M)) \xrightarrow{\;a\;} \operatorname{Ker}(\theta^i(M)) \longrightarrow 0.
$$

Whence $\operatorname{Ker}(\theta^i(M)) = 0$ by Nakayama's lemma. Thus we have $\theta^i(M)$ is an injection for all $i \in \mathbb{Z}$. Next we deduce a short exact sequence

$$
\operatorname{Cok}(\theta^i(M)) \xrightarrow{\;a\;} \operatorname{Cok}(\theta^i(M)) \longrightarrow 0.
$$

As above we conclude by Nakayama's lemma that $\operatorname{Cok}(\theta^i(M)) = 0$. $\square$

7.6 Complexes of flat modules

Throughout this section A denotes a ring. By a module is understood an A-module and by a complex is understood a complex of A-modules.

Recall that for complexes $X^{\boldsymbol{\cdot}}$ and $Y^{\boldsymbol{\cdot}}$ we define a complex $X^{\boldsymbol{\cdot}} \otimes_A Y^{\boldsymbol{\cdot}}$ by

$$
[X^{\boldsymbol{\cdot}} \otimes_A Y^{\boldsymbol{\cdot}}]^n = \bigoplus_{i+j=n} X^i \otimes_A Y^j
$$

$$
\partial^n(x^i \otimes y^j) = \partial^i(x^i) \otimes y^j + (-1)^i x^i \otimes \partial^j(y^j).
$$

Let us also recall that

$$
x^i \otimes y^j \mapsto (-1)^{ij} y^j \otimes x^i
$$

induces an isomorphism

$$
X^{\boldsymbol{\cdot}} \otimes_A Y^{\boldsymbol{\cdot}} \simeq Y^{\boldsymbol{\cdot}} \otimes_A X^{\boldsymbol{\cdot}}.
$$

Lemma 7.37. *Let $F^{\boldsymbol{\cdot}}$ be a bounded above complex of flat modules. If $X^{\boldsymbol{\cdot}}$ is any complex with $H^{\boldsymbol{\cdot}}(X^{\boldsymbol{\cdot}}) = 0$ then*

$$
H^{\boldsymbol{\cdot}}(X^{\boldsymbol{\cdot}} \otimes_A F^{\boldsymbol{\cdot}}) = 0.
$$

Proof. We fix $p \in \mathbb{Z}$ and then prove that $H^p(X^{\bullet} \otimes_A F^{\bullet}) = 0$ for all bounded above complexes with $H^{\bullet}(X^{\bullet}) = 0$. Let $X^{\bullet}$ have the form

$$\cdots \longrightarrow X^{n-1} \xrightarrow{\partial^{n-1}} X^n \xrightarrow{\partial^n} X^{n+1} \longrightarrow 0 \longrightarrow \cdots .$$

We can now form a short exact sequence of complexes

$$
\begin{array}{ccccccccc}
X_1^{\bullet} : & \longrightarrow & X^{n-1} & \longrightarrow & \mathrm{Ker}(\partial^n) & \longrightarrow & 0 & \longrightarrow & 0 \longrightarrow \\
& & \downarrow & & \downarrow & & \downarrow & & \downarrow \\
X^{\bullet} : & \longrightarrow & X^{n-1} & \longrightarrow & X^n & \longrightarrow & X^{n+1} & \longrightarrow & 0 \longrightarrow \\
& & \downarrow & & \downarrow & & \downarrow & & \downarrow \\
X_2^{\bullet} : & \longrightarrow & 0 & \longrightarrow & \mathrm{Im}(\partial^n) & \longrightarrow & X^{n+1} & \longrightarrow & 0 \longrightarrow
\end{array}
$$

tensor this sequence by $F^{\bullet}$ and consider the exact sequence

$$H^p(X_1^{\bullet} \otimes_A F^{\bullet}) \longrightarrow H^p(X^{\bullet} \otimes_A F^{\bullet}) \longrightarrow H^p(X_2^{\bullet} \otimes_A F^{\bullet})$$

and proceed by increasing induction on n In general a p-cocycle $\sum_i x^i \otimes f^{p-i}$ only depends on the truncation $X_1^{\bullet}$ for $n \gg 0$. $\square$

Proposition 7.38. *Let $F^{\bullet}$ be a bounded above complex of flat modules. Then for any quasi-isomorphism $f^{\bullet} : X^{\bullet} \to Y^{\bullet}$*

$$f^{\bullet} \otimes 1_{F^{\bullet}} : X^{\bullet} \otimes_A F^{\bullet} \to Y^{\bullet} \otimes_A F^{\bullet}$$

is a quasi-isomorphism.

Proof. Consider the mapping cone $C^{\bullet}(f^{\bullet})$ of $f^{\bullet}$ and note that $C^{\bullet}(f^{\bullet}) \otimes_A F^{\bullet}$ is the mapping cone of $f^{\bullet} \otimes 1_F^{\bullet}$. Conclusion by 7.37. $\square$

Proposition 7.39. *Let $f^{\bullet} : F^{\bullet} \to G^{\bullet}$ be a quasi-isomorphism of bounded above complexes of flat modules. Then for any complex $X^{\bullet}$*

$$1_{X^{\bullet}} \otimes f^{\bullet} : X^{\bullet} \otimes_A F^{\bullet} \to X^{\bullet} \otimes_A G^{\bullet}$$

is a quasi-isomorphism.

Proof. The same as that of 7.38. $\square$

Proposition 7.40. *Let A be a noetherian ring. For a bounded above complex $X^{\bullet}$ with finitely generated cohomology modules, there exists a bounded above complex $P^{\bullet}$ of finitely generated projective modules and a quasi-isomorphism $P^{\bullet} \to X^{\bullet}$.*

Proof. Use the procedure dual to the one used in the proof of 7.6, compare 7.7 and 7.12. □

Proposition 7.41. *Let A be a noetherian ring. For any bounded complex $F^{\cdot}$ of flat modules having finitely generated cohomology modules, there exists a bounded complex $P^{\cdot}$ of finitely generated projective modules and a quasi-isomorphism $P^{\cdot} \to F^{\cdot}$.*

If $F^i = 0$, $i < 0$ then we may choose $P^{\cdot}$ with $P^i = 0$, $i < 0$.

Proof. Assume to begin with that $F^i = 0$ for $i < 0$. Choose a bounded above complex of finitely generated projective modules and a quasi-isomorphism $g^{\cdot} : P^{\cdot} \to F^{\cdot}$. Let $Q^{\cdot}$ denote the complex

$$0 \longrightarrow \operatorname{Cok}(\partial^{-1}) \longrightarrow P^1 \longrightarrow P^2 \longrightarrow \cdots$$

and $f^{\cdot} : Q^{\cdot} \to F^{\cdot}$ the morphism induced by $g^{\cdot}$. This is a quasi-isomorphism, thus $H^{\cdot}(C^{\cdot}(f^{\cdot})) = 0$. Note $C^{-1}(f^{\cdot}) = Q^0$ and $C^i(f^{\cdot})$ is flat for $i \geq 0$. This shows that Q^0 has a finite resolution by flat modules, which implies that Q^0 is flat. But a finitely generated flat module over a noetherian ring is projective □

Lemma 7.42. *Let A be a noetherian ring with $\operatorname{Spec}(A)$ connected. If $P^{\cdot}$ and $Q^{\cdot}$ are bounded complexes of flat modules having finitely generated cohomology modules and such that $H^0(P^{\cdot} \otimes_A Q^{\cdot}) \simeq A$ and $H^i(P^{\cdot} \otimes_A Q^{\cdot}) = 0$, $i \neq 0$, then there exists a $n \in \mathbb{Z}$, a projective module of rank one L and a quasi-isomorphism $L[n] \to P^{\cdot}$.*

Proof. By 7.41 and 7.38 we may assume that $P^{\cdot}$ and $Q^{\cdot}$ are bounded complexes of finitely generated projective modules. Suppose first A is local with residue field k. We deduce a quasi-isomorphism

$$(P^{\cdot} \otimes_A k) \otimes_k (Q^{\cdot} \otimes_A k) \simeq k.$$

Thus we can find $p \in \mathbb{Z}$ such that $H^p(P^{\cdot} \otimes_A k) \neq 0$ and $H^i(P^{\cdot} \otimes_A k) = 0$ for $i \neq p$. Use 2.34 to deduce from this that $P^{\cdot} \simeq A^m[-p]$ for some m. Similarly we can find n, q such that $Q^{\cdot} \simeq A^n[-q]$. It now follows that $m = n = 1$ and $p + q = 0$.

In the general case consider the function $r : \operatorname{Spec}(A) \to \mathbb{Z}$ characterized by $H^{r(\mathfrak{p})}(k_{\mathfrak{p}} \otimes_{A_{\mathfrak{p}}} P_{\mathfrak{p}}^{\cdot}) \neq 0$. Our previous investigation shows that r is locally constant. Now $\operatorname{Spec}(A)$ is connected and therefore r is constant, say $r = p$. It follows that $H^i(P^{\cdot}) = 0$ for $i \neq p$ and that $H^p(P^{\cdot}) = L$ is a projective module of rank one. An easy homotopy argument shows that $P^{\cdot} \simeq L[-p]$. □

7.7 Generalized evaluation maps

Let A denote a ring. For complexes $X^{\cdot}, Y^{\cdot}, Z^{\cdot}$ of A-modules we have, $m, n \in \mathbb{Z}$

$$\operatorname{Hom}_A^n(Y^{\cdot}, Z^{\cdot}) \otimes_A \operatorname{Hom}_A^m(X^{\cdot}, Y^{\cdot}) \to \operatorname{Hom}_A^{m+n}(X^{\cdot}, Z^{\cdot})$$

given by $(g, f) \to g \circ f$. Note

$$\partial(g \circ f) = \partial g \circ f + (-1)^n g \circ \partial f$$

thus this gives rise to the *composition* map

$$\operatorname{Hom}_A^{\cdot}(Y^{\cdot}, Z^{\cdot}) \otimes_A \operatorname{Hom}_A^{\cdot}(X^{\cdot}, Y^{\cdot}) \to \operatorname{Hom}_A^{\cdot}(X^{\cdot}, Z^{\cdot}). \qquad (7.1)$$

Let now $X^{\cdot}, E^{\cdot}, F^{\cdot}$ be complexes and consider the evaluation map

$$ev : X^{\cdot} \to \operatorname{Hom}_A^{\cdot}(\operatorname{Hom}_A^{\cdot}(X^{\cdot}, E^{\cdot}), E^{\cdot})$$

tensor this with $\operatorname{Hom}_A^{\cdot}(E^{\cdot}, F^{\cdot})$ to obtain a map

$$\operatorname{Hom}_A^{\cdot}(E^{\cdot}, F^{\cdot}) \otimes_A X^{\cdot} \to \operatorname{Hom}_A^{\cdot}(E^{\cdot}, F^{\cdot}) \otimes_A \operatorname{Hom}_A^{\cdot}(\operatorname{Hom}_A^{\cdot}(X^{\cdot}, E^{\cdot}), E^{\cdot}).$$

Compose this map with the composition map

$$\operatorname{Hom}_A^{\cdot}(E^{\cdot}, F^{\cdot}) \otimes_A \operatorname{Hom}_A^{\cdot}(\operatorname{Hom}_A^{\cdot}(X^{\cdot}, E^{\cdot}), E^{\cdot}) \to \operatorname{Hom}_A^{\cdot}(\operatorname{Hom}_A^{\cdot}(X^{\cdot}, E^{\cdot}), F^{\cdot})$$

to obtain the *generalized evaluation* map

$$\operatorname{Hom}_A^{\cdot}(E^{\cdot}, F^{\cdot}) \otimes_A X^{\cdot} \to \operatorname{Hom}_A^{\cdot}(\operatorname{Hom}_A^{\cdot}(X^{\cdot}, E^{\cdot}), F^{\cdot}). \qquad (7.2)$$

The canonical map

$$A \to \operatorname{Hom}_A^{\cdot}(E^{\cdot}, E^{\cdot})$$

gives rise to

$$\operatorname{Hom}_A^{\cdot}(\operatorname{Hom}_A^{\cdot}(E^{\cdot}, E^{\cdot}), F^{\cdot}) \to F^{\cdot}.$$

Compose this map with (7.2) to obtain a map

$$\operatorname{Hom}_A^{\cdot}(E^{\cdot}, F^{\cdot}) \otimes_A E^{\cdot} \to F^{\cdot}. \qquad (7.3)$$

Proposition 7.43. *Suppose A is a noetherian ring and let $E^{\cdot}$ be a bounded below complex of injective modules and $F^{\cdot}$ a bounded complex of injective modules. Then $\operatorname{Hom}_A^{\cdot}(E^{\cdot}, F^{\cdot})$ is a bounded above complex of flat modules,* *and for any bounded above complex $X^{\cdot}$ with finitely generated cohomology, the generalized evaluation map*

$$\operatorname{Hom}_A^{\cdot}(E^{\cdot}, F^{\cdot}) \otimes_A X^{\cdot} \to \operatorname{Hom}_A^{\cdot}(\operatorname{Hom}_A^{\cdot}(X^{\cdot}, E^{\cdot}), F^{\cdot})$$

is a quasi-isomorphism.

Proof. For the first part consider injective modules E and F and a finitely generated module M. The generalized evaluation map

$$\operatorname{Hom}_A(E, F) \otimes_A M \to \operatorname{Hom}_A(\operatorname{Hom}_A(M, E), F)$$

is an isomorphism, as one sees by considering a presentation of M, $A^n \to A^m \to M \to 0$. It follows that $\operatorname{Hom}_A(E, F)$ is a flat module. To prove the second part one proceeds by the methods developed in 7.14. ☐

Corollary 7.44. *Let $E^{\cdot}$ be a dualizing complex for A and $F^{\cdot}$ a bounded complex of injective modules with finitely generated cohomology. Then the canonical map (7.3)*

$$\operatorname{Hom}_A^{\cdot}(E^{\cdot}, F^{\cdot}) \otimes_A E^{\cdot} \to F^{\cdot}$$

and the composition map (7.1)

$$\operatorname{Hom}_A^{\cdot}(E^{\cdot}, F^{\cdot}) \otimes_A \operatorname{Hom}_A^{\cdot}(F^{\cdot}, E^{\cdot}) \to \operatorname{Hom}_A^{\cdot}(F^{\cdot}, F^{\cdot})$$

are both quasi-isomorphisms

Proof. We have quasi-isomorphisms 7.43

$$\operatorname{Hom}_A^{\cdot}(E^{\cdot}, F^{\cdot}) \otimes_A E^{\cdot} \to \operatorname{Hom}_A^{\cdot}(\operatorname{Hom}_A^{\cdot}(E^{\cdot}, E^{\cdot}), F^{\cdot})$$

and

$$A \to \operatorname{Hom}_A^{\cdot}(E^{\cdot}, E^{\cdot})$$

thus the first assertion follows from 7.3.

For the second consider the commutative diagram:

$$
\begin{array}{ccc}
\operatorname{Hom}_A^{\cdot}(E^{\cdot}, F^{\cdot}) \otimes_A \operatorname{Hom}_A^{\cdot}(F^{\cdot}, E^{\cdot}) & \longrightarrow & \operatorname{Hom}_A^{\cdot}(F^{\cdot}, F^{\cdot}) \\
\downarrow & & \| \\
\operatorname{Hom}_A^{\cdot}(\operatorname{Hom}_A^{\cdot}(\operatorname{Hom}_A^{\cdot}(F^{\cdot}, E^{\cdot}), E^{\cdot}), F^{\cdot}) & \longrightarrow & \operatorname{Hom}_A^{\cdot}(F^{\cdot}, F^{\cdot})
\end{array}
$$

where the horizontal map is a quasi-isomorphism from 7.43 in the special case $X^{\cdot} = \operatorname{Hom}_A^{\cdot}(F^{\cdot}, E^{\cdot})$ and the right vertical arrow is obtained by applying $\operatorname{Hom}_A^{\cdot}(-, F^{\cdot})$ to

$$ev : F^{\cdot} \to \operatorname{Hom}_A^{\cdot}(\operatorname{Hom}_A^{\cdot}(F^{\cdot}, E^{\cdot}), E^{\cdot}). \qquad \qquad ☐$$

7.8 Uniqueness of dualizing complexes

We have the fundamental uniqueness theorem.

Theorem 7.45. *Let A be a noetherian ring with $\mathrm{Spec}(A)$ connected. If $E^{\cdot}$ and $F^{\cdot}$ are dualizing complexes, then there exists $n \in \mathbb{Z}$ and a projective module L of rank one such that $E^{\cdot} \otimes_A L[n]$ is homotopy equivalent to $F^{\cdot}$.*

Proof. $\mathrm{Hom}_A^{\cdot}(E^{\cdot}, F^{\cdot})$ and $\mathrm{Hom}_A^{\cdot}(F^{\cdot}, E^{\cdot})$ are bounded complexes of flat modules, 7.43, with finitely generated cohomology modules and by 7.44 we have a quasi-isomorphism

$$\mathrm{Hom}_A^{\cdot}(E^{\cdot}, F^{\cdot}) \otimes_A \mathrm{Hom}_A^{\cdot}(F^{\cdot}, E^{\cdot}) \to \mathrm{Hom}_A^{\cdot}(F^{\cdot}, F^{\cdot}).$$

It follows that

$$H^i(\mathrm{Hom}_A^{\cdot}(E^{\cdot}, F^{\cdot}) \otimes_A \mathrm{Hom}_A^{\cdot}(F^{\cdot}, E^{\cdot})) \simeq \begin{cases} A, & i = 0 \\ 0, & i \neq 0. \end{cases}$$

Thus by 7.42 we can find $n \in \mathbb{Z}$, a projective module of rank one L and a quasi-isomorphism

$$L[n] \to \mathrm{Hom}_A^{\cdot}(E^{\cdot}, F^{\cdot}).$$

Tensor this with $E^{\cdot}$ to obtain a quasi-isomorphism, 7.38

$$L[n] \otimes_A E^{\cdot} \to \mathrm{Hom}_A^{\cdot}(E^{\cdot}, F^{\cdot}) \otimes_A E^{\cdot}.$$

Compose this with the quasi-isomorphism 7.44

$$\mathrm{Hom}_A^{\cdot}(E^{\cdot}, F^{\cdot}) \otimes_A E^{\cdot} \to F^{\cdot}$$

to obtain a quasi-isomorphism

$$L[n] \otimes_A E^{\cdot} \to F^{\cdot}.$$

This is a homotopy equivalence by 7.5. $\qquad\square$

Corollary 7.46. *Let A be a noetherian local ring. If $E^{\cdot}$ and $F^{\cdot}$ are dualizing complexes, then there exists $n \in \mathbb{Z}$ such that $E^{\cdot}[n]$ is homotopy equivalent to $F^{\cdot}$.*

Proof. L in 7.45 is a free module. $\qquad\square$

Chapter 8

Local Duality

Let A be a ring. By a module is understood an A-module and by a complex is understood a complex of A-modules.

Given a complex $E^{\cdot}$ of injective modules and a module M then for all $i \in \mathbb{Z}$ we put

$$\mathrm{Ext}_A^i(M, E^{\cdot}) = H^i(\mathrm{Hom}_A^{\cdot}(M, E^{\cdot})).$$

For a bounded below complex $X^{\cdot}$ we choose a quasi-isomorphism $X^{\cdot} \to E^{\cdot}$ into a bounded below complex of injective modules, 7.6, and define

$$\mathrm{Ext}_A^i(M, X^{\cdot}) = \mathrm{Ext}_A^i(M, E^{\cdot}).$$

By 7.4 this is independent of choice.

Given a complex $P^{\cdot}$ of projective modules and a module M then for all $i \in \mathbb{Z}$ we put

$$\mathrm{Tor}_i^A(M, P^{\cdot}) = H^{-i}(M \otimes_A P^{\cdot}).$$

For a bounded above complex $X^{\cdot}$ we choose a quasi-isomorphism $P^{\cdot} \to X^{\cdot}$ from a bounded above complex of projective modules, 7.40, and define

$$\mathrm{Tor}_i^A(M, X^{\cdot}) = \mathrm{Tor}_i^A(M, P^{\cdot}).$$

By 7.12 this is independent of choice.

8.1 Poincaré series

Throughout this section A denotes a noetherian local ring with maximal ideal $\mathfrak{m}$ and residue field k.

Definition 8.1. Let $X^{\bullet}$ be a complex with finitely generated cohomology modules. If $X^{\bullet}$ is bounded below define the *Poincaré series*

$$\mu(X^{\bullet}, t) = \sum_i \operatorname{rank}_k(\operatorname{Ext}^i_A(k, X^{\bullet}))t^i$$

and if $X^{\bullet}$ is bounded above,

$$\beta(X^{\bullet}, t) = \sum_i \operatorname{rank}_k(\operatorname{Tor}^A_i(k, X^{\bullet}))t^i.$$

If $X^{\bullet}$ is a finitely generated module, then these series are the ordinary Poincaré series.

Proposition 8.2. *Let $X^{\bullet}$ be a complex with finitely generated cohomology modules and assume $H^{\bullet}(X^{\bullet}) \neq 0$. If $X^{\bullet}$ is bounded below then $\mu(X^{\bullet}, t) \neq 0$. If $X^{\bullet}$ is bounded above then $\beta(X^{\bullet}, t) \neq 0$.*

Proof. Suppose $X^{\bullet}$ is bounded below, and let $i \in \mathbb{Z}$ be such that $H^i(X^{\bullet}) \neq 0$ and $H^j(X^{\bullet}) = 0$, $j < i$. Choose $\mathfrak{p} \in \operatorname{Ass}(H^i(X^{\bullet}))$. It follows that $\operatorname{Ext}^i_A(k_{\mathfrak{p}}, X^{\bullet}_{\mathfrak{p}}) \neq 0$. Choose a saturated chain of prime ideals between $\mathfrak{p}$ and $\mathfrak{m}$ and apply 7.16 to see that $\operatorname{Ext}^{\bullet}_A(k, X^{\bullet}) \neq 0$. The second part follows from Nakayama's lemma. $\square$

Proposition 8.3. *Let $X^{\bullet}$ be a bounded above complex with finitely generated cohomology modules and $E^{\bullet}$ a bounded below complex of injective modules with finitely generated cohomology modules. Then*

$$\mu(\operatorname{Hom}^{\bullet}_A(X^{\bullet}, E^{\bullet}), t) = \beta(X^{\bullet}, t)\mu(E^{\bullet}, t).$$

If $X^{\bullet}$ and $E^{\bullet}$ are bounded, then

$$\beta(\operatorname{Hom}^{\bullet}_A(X^{\bullet}, E^{\bullet}), t) = \mu(X^{\bullet}, t)\mu(E^{\bullet}, t^{-1}).$$

Proof. To prove the first formula we may assume by 7.40 that $X^{\bullet}$ is a complex of finitely generated free modules. We have isomorphisms of complexes

$$\operatorname{Hom}^{\bullet}_A(k, \operatorname{Hom}^{\bullet}_A(X^{\bullet}, E^{\bullet})) \simeq \operatorname{Hom}^{\bullet}_A(k \otimes_A X^{\bullet}, E^{\bullet})$$
$$\simeq \operatorname{Hom}^{\bullet}_A(k \otimes_A X^{\bullet}, \operatorname{Hom}^{\bullet}_A(k, E^{\bullet}))$$

from which the first formula follows.

To prove the second formula choose a bounded below complex $F^{\bullet}$ of injective modules and a quasi-isomorphism $X^{\bullet} \to F^{\bullet}$. By 7.43 $\operatorname{Hom}^{\bullet}_A(F^{\bullet}, E^{\bullet})$

is a bounded above complex of flat modules. If $L^{\cdot} \to k$ is a free resolution we get a quasi-isomorphism, 7.38,

$$\operatorname{Hom}_A^{\cdot}(F^{\cdot}, E^{\cdot}) \otimes_A L^{\cdot} \to \operatorname{Hom}_A^{\cdot}(F^{\cdot}, E^{\cdot}) \otimes_A k$$

and again by 7.43 a quasi-isomorphism

$$\operatorname{Hom}_A^{\cdot}(F^{\cdot}, E^{\cdot}) \otimes_A k \to \operatorname{Hom}_A^{\cdot}(\operatorname{Hom}_A^{\cdot}(k, F^{\cdot}), E^{\cdot}).$$

It is now easy to conclude by the elementary isomorphism

$$\operatorname{Hom}_A^{\cdot}(\operatorname{Hom}_A^{\cdot}(k, F^{\cdot}), E^{\cdot}) \simeq \operatorname{Hom}_A^{\cdot}(\operatorname{Hom}_A^{\cdot}(k, F^{\cdot}), \operatorname{Hom}_A^{\cdot}(k, E^{\cdot})). \qquad \square$$

Corollary 8.4. *Let $X^{\cdot}$ be a bounded above complex with finitely generated cohomology modules and $E^{\cdot}$ a bounded below complex of injective modules with finitely generated cohomology modules. If $H^{\cdot}(X^{\cdot}) \neq 0$ and $H^{\cdot}(E^{\cdot}) \neq 0$ then*

$$H^{\cdot}(\operatorname{Hom}_A^{\cdot}(X^{\cdot}, E^{\cdot})) \neq 0.$$

Proof. From 8.2 follows that $\beta(X^{\cdot}, t) \neq 0$ and $\mu(E^{\cdot}, t) \neq 0$ thus we have $\mu(\operatorname{Hom}_A^{\cdot}(X^{\cdot}, E^{\cdot}), t) \neq 0$ by the first formula in 8.3. $\qquad \square$

Definition 8.5. A complex $X^{\cdot}$ is said to have *finite injective amplitude* if $X^{\cdot}$ is bounded below and there exists a quasi-isomorphism $X^{\cdot} \to E^{\cdot}$ where $E^{\cdot}$ is a bounded complex of injective modules.

The complex $X^{\cdot}$ is said to have *finite projective amplitude* if $X^{\cdot}$ is bounded above and there exists a quasi-isomorphism $P^{\cdot} \to X^{\cdot}$ where $P^{\cdot}$ is a bounded complex of projective modules.

Let $X^{\cdot}$ be a complex with finitely generated cohomology modules. It is clear that if $X^{\cdot}$ has finite injective amplitude resp. finite projective amplitude, then $\mu(X^{\cdot}, t)$ resp. $\beta(X^{\cdot}, t)$ is a finite series.

We have the following converse.

Proposition 8.6. *Let $X^{\cdot}$ be a complex with finitely generated cohomology modules. If $X^{\cdot}$ is bounded below and the series $\mu(X^{\cdot}, t)$ is finite, then $X^{\cdot}$ has finite injective amplitude.*

Similarly if $X^{\cdot}$ is bounded above and the series $\beta(X^{\cdot}, t)$ is finite, then $X^{\cdot}$ has finite projective amplitude.

Proof. The first part is a reformulation of 7.17.

For the second part we can assume that $X^{\cdot} = L^{\cdot}$ is a bounded above complex of finitely generated free modules. The proof of 2.35 shows that

$$H^i(L^{\cdot} \otimes_A k) = 0 \quad \Rightarrow \quad H^i(L^{\cdot}) = 0$$

and that $\mathrm{Ker}(\partial^i)$ is a direct summand in L^i. From this the assertion easily follows.   □

Remark 8.7. Suppose $D^{\cdot}$ is a dualizing complex for A. The μ-series for $D^{\cdot}$ has the form, 7.30,

$$\mu(D^{\cdot}, t) = t^r, \qquad r \in \mathbb{Z}.$$

For a bounded complex $X^{\cdot}$ with finitely generated cohomology it follows from 8.3 that

$$\mu(\mathrm{Hom}_A^{\cdot}(X^{\cdot}, D^{\cdot}), t) = t^r \beta(X^{\cdot}, t)$$
$$\beta(\mathrm{Hom}_A^{\cdot}(X^{\cdot}, D^{\cdot}), t) = t^{-r} \mu(X^{\cdot}, t).$$

In particular, if $X^{\cdot}$ has finite injective amplitude then $\mathrm{Hom}_A^{\cdot}(X^{\cdot}, D^{\cdot})$ has finite projective amplitude and conversely.

Proposition 8.8. *Let $L^{\cdot}$ be a bounded complex of finitely generated free modules and $X^{\cdot}$ a bounded below complex with finitely generated cohomology modules. Then*

$$\mu(L^{\cdot} \otimes_A X^{\cdot}, t) = \beta(L^{\cdot}, t^{-1})\mu(X^{\cdot}, t).$$

Proof. Choose a quasi-isomorphism $X^{\cdot} \to E^{\cdot}$ where $E^{\cdot}$ is a bounded below complex of injective modules. Then $L^{\cdot} \otimes_A X^{\cdot} \to L^{\cdot} \otimes_A E^{\cdot}$ is a quasi-isomorphism as it follows by 7.38, thus it suffices to prove

$$\mu(L^{\cdot} \otimes_A E^{\cdot}, t) = \beta(L^{\cdot}, t^{-1})\mu(E^{\cdot}, t).$$

We have an isomorphism of complexes $\mathrm{Hom}_A^{\cdot}(\mathrm{Hom}_A^{\cdot}(L^{\cdot}, A), E^{\cdot}) \simeq L^{\cdot} \otimes_A E^{\cdot}$, thus by the first formula in 8.3

$$\mu(L^{\cdot} \otimes_A E^{\cdot}, t) = \beta(\mathrm{Hom}_A^{\cdot}(L^{\cdot}, A), t)\mu(E^{\cdot}, t).$$

We can now conclude by remarking

$$\beta(\mathrm{Hom}_A^{\cdot}(L^{\cdot}, A), t) = \beta(L^{\cdot}, t^{-1}).$$ □

Proposition 8.9. *Suppose there exists a bounded complex $X^{\cdot}$ with finitely generated cohomology modules and $H^{\cdot}(X^{\cdot}) \neq 0$, which has finite injective amplitude and finite projective amplitude. Then A is Gorenstein.*

Proof. By 7.41 we can find a quasi-isomorphism $L^{\cdot} \to X^{\cdot}$ where $L^{\cdot}$ is a bounded complex of finitely generated free modules. By 8.8 we have $\mu(L^{\cdot}, t) = \beta(L^{\cdot}, t^{-1})\mu(A, t)$ and whence

$$\mu(X^{\cdot}, t) = \beta(X^{\cdot}, t^{-1})\mu(A, t). \tag{8.1}$$

From the last formula in 8.3 applied to A and $E^{\cdot}$ where $E^{\cdot}$ is a bounded complex of injective modules quasi-isomorphic to $X^{\cdot}$ we get

$$\beta(X^{\cdot}, t) = \mu(X^{\cdot}, t^{-1})\mu(A, t). \tag{8.2}$$

Combining (8.1) and (8.2) we get

$$\mu(X^{\cdot}, t) = \mu(X^{\cdot}, t)\mu(A, t^{-1})\mu(A, t).$$

Since $\mu(X^{\cdot}, t)$ is non trivial we get

$$\mu(A, t^{-1})\mu(A, t) = 1.$$

From this one deduces easily that $\mu(A, t)$ has the form

$$\mu(A, t) = t^r, \qquad r \in \mathbb{N}$$

which forces A to be Gorenstein. $\qquad\qquad\square$

8.2   Grothendieck's local duality theorem

Let A denote a noetherian local ring with maximal ideal $\mathfrak{m}$ and residue field k. By a module we understand an A-module and by a complex we understand a complex of A-modules.

For a bounded below complex $X^{\cdot}$, we define the *local cohomology complex*

$$R^{\cdot}\Gamma_{\mathfrak{m}}(X^{\cdot})$$

as the complex obtained in the following way: choose a quasi-isomorphism $X^{\cdot} \to E^{\cdot}$ where $E^{\cdot}$ is a bounded below complex of injective modules, 7.6, and put $R^{\bullet}\Gamma_{\mathfrak{m}}(X^{\bullet}) = \Gamma_{\mathfrak{m}}(X^{\bullet})$. This complex is unique up to homotopy, 7.4, and we define

$$R^i\Gamma_{\mathfrak{m}}(X^{\bullet}) = H^i(R^{\bullet}\Gamma_{\mathfrak{m}}(X^{\bullet})).$$

In particular for a module M (considered as a complex) we have with the notation of section 6.1

$$R^i\Gamma_{\mathfrak{m}}(M) = H^i_{\mathfrak{m}}(M).$$

Definition 8.10. A dualizing complex $D^{\bullet}$ for A is called *normalized* if

$$\operatorname{Ext}^d_A(k, D^{\bullet}) \simeq k, \qquad d = \dim(A).$$

Note by the results of section 7.5 there is only one nonzero Ext-group being isomorphic to k.

Lemma 8.11. *Let $D^{\bullet}$ be a normalized dualizing complex for A. Then if E is an injective envelope of k, we have*

$$R^i\Gamma_{\mathfrak{m}}(D^{\bullet}) \simeq \begin{cases} 0, & i \neq \dim(A) \\ E, & i = \dim(A). \end{cases}$$

Proof. We may assume that $D^{\bullet}$ is a minimal complex. Conclusion by 7.34. $\square$

In the rest of this section, $D^{\bullet}$ denotes a normalized dualizing complex, $d = \dim(A)$ and the exponential $-^{\vee}$ will denote the Matlis dual. We are going to construct for any complex $X^{\bullet}$ a map of complexes

Definition 8.12.

$$\Gamma_{\mathfrak{m}}(X^{\bullet})[d] \to \operatorname{Hom}^{\bullet}_A(X^{\bullet}, D^{\bullet})^{\vee}.$$

Let us first notice that $\Gamma_{\mathfrak{m}}$ give rise to a map of complexes

$$\operatorname{Hom}^{\bullet}_A(X^{\bullet}, D^{\bullet}) \to \operatorname{Hom}^{\bullet}_A(\Gamma_{\mathfrak{m}}(X^{\bullet}), \Gamma_{\mathfrak{m}}(D^{\bullet})).$$

Apply $\operatorname{Hom}^{\bullet}_A(-, \Gamma_{\mathfrak{m}}(D^{\bullet}))$ to the arrow above to get

$$\operatorname{Hom}^{\bullet}_A(\operatorname{Hom}^{\bullet}_A(\Gamma_{\mathfrak{m}}(X^{\bullet}), \Gamma_{\mathfrak{m}}(D^{\bullet})), \Gamma_{\mathfrak{m}}(D^{\bullet}))$$
$$\to \operatorname{Hom}^{\bullet}_A(\operatorname{Hom}^{\bullet}_A(X^{\bullet}, D^{\bullet}), \Gamma_{\mathfrak{m}}(D^{\bullet}))$$

compose this with the evaluation map

$$\Gamma_{\mathfrak{m}}(X^{\bullet}) \to \operatorname{Hom}^{\bullet}_A(\operatorname{Hom}^{\bullet}_A(\Gamma_{\mathfrak{m}}(X^{\bullet}), \Gamma_{\mathfrak{m}}(D^{\bullet})), \Gamma_{\mathfrak{m}}(D^{\bullet}))$$

to obtain a map

$$\Gamma_{\mathfrak{m}}(X^{\textbf{.}}) \to \operatorname{Hom}_A^{\textbf{.}}(\operatorname{Hom}_A^{\textbf{.}}(\operatorname{Hom}_A^{\textbf{.}}(X^{\textbf{.}}, D^{\textbf{.}}), \Gamma_{\mathfrak{m}}(D^{\textbf{.}})).$$

Finally use a homotopy equivalence 8.11

$$\Gamma_{\mathfrak{m}}(D^{\textbf{.}}) \simeq E[-d]$$

to get the desired map

$$\Gamma_{\mathfrak{m}}(X^{\textbf{.}})[d] \to \operatorname{Hom}_A^{\textbf{.}}(X^{\textbf{.}}, D^{\textbf{.}})^{\vee}.$$

In particular if $X^{\textbf{.}}$ is an injective resolution of a module M, we get a map

$$R^i\Gamma_{\mathfrak{m}}(M) \to \operatorname{Ext}_A^{d-i}(M, D^{\textbf{.}})^{\vee}, \qquad i \in \mathbb{Z}.$$

Theorem 8.13 (Grothendieck's local duality theorem). *Let M be a finitely generated module. Then for $i \in \mathbb{Z}$ the canonical map*

$$R^i\Gamma_{\mathfrak{m}}(M) \to \operatorname{Ext}_A^{d-i}(M, D^{\textbf{.}})^{\vee}$$

is an isomorphism.

Proof. Let us first remark that if $X^{\textbf{.}}$ is a bounded below complex with finitely generated cohomology modules, and such that $\Gamma_{\mathfrak{m}}(X^{\textbf{.}}) \simeq X^{\textbf{.}}$, then the canonical map 8.12 is a quasi-isomorphism as the reader easily checks from the definition. In particular if $X^{\textbf{.}}$ is a minimal injective resolution of k. Thus the theorem is true for $M = k$. The general case now follows by the method used in the proof of 7.36. $\qquad\square$

Corollary 8.14. *Let $X^{\textbf{.}}$ be a bounded below complex with finitely generated cohomology modules. Then the canonical map*

$$R^{\textbf{.}}\Gamma_{\mathfrak{m}}(X^{\textbf{.}}) \to \operatorname{Hom}_A^{\textbf{.}}(X^{\textbf{.}}, D^{\textbf{.}})^{\vee}$$

is a quasi-isomorphism.

Proof. Let us first remark that the conclusion is true whenever $X^{\textbf{.}}$ has only one nonzero cohomology module as it follows from 8.4 and the remark that our map is compatible with decalage. In the general case suppose $H^i(X^{\textbf{.}}) = 0$ for $i < n$. Let $Z^{\textbf{.}}$ denote an injective resolution of $H^n(X^{\textbf{.}})$ and consider a map $f : Z^{\textbf{.}}[-n] \to X^{\textbf{.}}$ such that $H^n(f^{\textbf{.}})$ is an isomorphism. Given a fixed $p \in \mathbb{Z}$ and suppose that we know that

$$H^q(\Gamma_{\mathfrak{m}}(Y^{\textbf{.}})[d]) \simeq H^q(\operatorname{Hom}_A^{\textbf{.}}(Y^{\textbf{.}}, D^{\textbf{.}})^{\vee})$$

for $q \leq p$ and for all complexes $Y^{\cdot}$ as above with $H^i(Y^{\cdot}) = 0$, $i < n + 1$. Applying this to the mapping cone of $f : Z^{\cdot}[-n] \to X^{\cdot}$ we would conclude by the five lemma that

$$H^q(\Gamma_{\mathfrak{m}}(X^{\cdot})[d]) \simeq H^q(\operatorname{Hom}_A^{\cdot}(X^{\cdot}, D^{\cdot})^{\vee})$$

for $q \leq p$. We leave it to the reader to arrange a decreasing induction on n with fixed p.   □

Remark 8.15. We shall here give a second proof of the local duality theorem. Remark first that for any bounded above complex $L^{\cdot}$ of finitely generated free modules and any bounded below complex of injective modules $E^{\cdot}$ we have an isomorphism of complexes

$$\Gamma_{\mathfrak{m}}(\operatorname{Hom}_A^{\cdot}(L^{\cdot}, E^{\cdot})) \simeq \operatorname{Hom}_A^{\cdot}(L^{\cdot}, \Gamma_{\mathfrak{m}}(E^{\cdot})).$$

If we now start with a bounded above complex $X^{\cdot}$ with finitely generated cohomology modules we can choose a quasi-isomorphism $L^{\cdot} \to X^{\cdot}$ where $L^{\cdot}$ is as above. This induces a quasi-isomorphism

$$\operatorname{Hom}_A^{\cdot}(X^{\cdot}, E^{\cdot}) \simeq \operatorname{Hom}_A^{\cdot}(L^{\cdot}, E^{\cdot}).$$

Notice that the complex to the right consists of injective modules, whence by definition

$$R^{\cdot}\Gamma_{\mathfrak{m}}(\operatorname{Hom}_A^{\cdot}(X^{\cdot}, E^{\cdot})) \simeq \Gamma_{\mathfrak{m}}(\operatorname{Hom}_A^{\cdot}(L^{\cdot}, E^{\cdot})).$$

Note also that the quasi-isomorphism $L^{\cdot} \to X^{\cdot}$ induces a quasi-isomorphism

$$\operatorname{Hom}_A^{\cdot}(X^{\cdot}, \Gamma_{\mathfrak{m}}(E^{\cdot})) \to \operatorname{Hom}_A^{\cdot}(L^{\cdot}, \Gamma_{\mathfrak{m}}(E^{\cdot}))$$

here we have used that $\Gamma_{\mathfrak{m}}(E^{\cdot})$ consists of injective modules. Collecting this together we have obtained a quasi-isomorphism

$$R^{\cdot}\Gamma_{\mathfrak{m}}(\operatorname{Hom}_A^{\cdot}(X^{\cdot}, E^{\cdot})) \simeq \operatorname{Hom}_A^{\cdot}(X^{\cdot}, \Gamma_{\mathfrak{m}}(E^{\cdot})).$$

In particular we get, if $E^{\cdot} = D^{\cdot}$ is a normalized dualizing complex, $\Gamma_{\mathfrak{m}}(D^{\cdot}) \simeq E[-d]$ and therefore

$$R^{\cdot}\Gamma_{\mathfrak{m}}(\operatorname{Hom}_A^{\cdot}(X^{\cdot}, D^{\cdot})) \simeq X^{\cdot \vee}[-d].$$

If now $Y^{\cdot}$ is a bounded below complex with finitely generated cohomology modules we get with $X^{\cdot} = \operatorname{Hom}_A^{\cdot}(Y^{\cdot}, D^{\cdot})$

$$R^{\cdot}\Gamma_{\mathfrak{m}}(\operatorname{Hom}_A^{\cdot}(\operatorname{Hom}_A^{\cdot}(Y^{\cdot}, D^{\cdot}), D^{\cdot})) \simeq \operatorname{Hom}_A(Y^{\cdot}, D^{\cdot})^{\vee}[-d],$$

and using the quasi-isomorphisms

$$ev : Y^{\cdot} \to \mathrm{Hom}_A^{\cdot}(\mathrm{Hom}_A^{\cdot}(Y^{\cdot}, D^{\cdot}), D^{\cdot}),$$
$$R^{\cdot}\Gamma_{\mathfrak{m}}(Y^{\cdot}) \simeq \mathrm{Hom}_A(Y^{\cdot}, D^{\cdot})^{\vee}[-d].$$

8.3 Duality for Cohen–Macaulay modules

In this section we shall use the local duality theorem 8.4 to study duality for Cohen–Macaulay modules. Let A denote a noetherian local ring of dimension d and $D^{\cdot}$ a normalized dualizing complex for A.

Theorem 8.16. *Let $N \neq 0$ denote a finitely generated A-module of dimension n. Then*

(1) $\mathrm{Ext}_A^{d-n}(N, D^{\cdot}) \neq 0$.

(2) $\mathrm{Ext}_A^{d-n}(N, D^{\cdot})$ *is the only nonzero cohomology module of the complex* $\mathrm{Hom}_A^{\cdot}(N, D^{\cdot})$ *if and only if N is a Cohen–Macaulay module.*

Proof. By 8.13 we have for $i \in \mathbb{Z}$

$$R^i\Gamma_{\mathfrak{m}}(N) \simeq \mathrm{Ext}_A^{n-d}(N, D^{\cdot})^{\vee}$$

thus (1) follows from 6.12, and (2) from 6.7 and 6.8. $\qquad\square$

Theorem 8.17. *Suppose $N \neq 0$ is a finitely generated Cohen–Macaulay module of dimension n. Then*

(1) $\mathrm{Ext}_A^{d-n}(N, D^{\cdot})$ *is a Cohen–Macaulay module of dimension n.*

(2) $N \simeq \mathrm{Ext}_A^{d-n}(\mathrm{Ext}_A^{d-n}(N, D^{\cdot}), D^{\cdot})$.

(3) $\mu(N, t) = t^n \beta(\mathrm{Ext}_A^{d-n}(N, D^{\cdot}), t)$
$\beta(N, t) = t^{-n} \mu(\mathrm{Ext}_A^{d-n}(N, D^{\cdot}), t)$.

Proof. By 8.16 we have a quasi-isomorphism

$$\mathrm{Ext}_A^{d-n}(N, D^{\cdot})[n - d] \simeq \mathrm{Hom}_A^{\cdot}(N, D^{\cdot}).$$

Thus by duality a quasi-isomorphism

$$\mathrm{Hom}_A^{\cdot}(\mathrm{Ext}_A^{d-n}(N, D^{\cdot}), D^{\cdot}) \to N[n - d].$$

This proves (2). Applying the criterion 8.16(2) to the module $\mathrm{Ext}_A^{d-n}(N, D^{\cdot})$ we see that this module is Cohen–Macaulay. To prove (3) let us first note that $\mu(D^{\cdot}, t) = t^d$. From 8.7 follows

$$\mu(\mathrm{Hom}_A^{\cdot}(N, D^{\cdot}), t) = t^d \beta(N, t)$$
$$\beta(\mathrm{Hom}_A^{\cdot}(N, D^{\cdot}), t) = t^{-d} \mu(N, t).$$

We have general formulas for decalage

$$\mu(Y^{\cdot}[p], t) = t^{-p} \mu(Y^{\cdot}, t)$$
$$\beta(Y^{\cdot}[p], t) = t^p \beta(Y^{\cdot}, t).$$

Now the result follows from the quasi-isomorphism

$$\mathrm{Ext}_A^{d-n}(N, D^{\cdot})[n - d] \simeq \mathrm{Hom}_A^{\cdot}(N, D^{\cdot}). \qquad \square$$

Proposition 8.18. *Suppose the ring A has a dualizing complex. Then the sets*

$$\{\mathfrak{p} \in \mathrm{Spec}(A) \mid A_{\mathfrak{p}} \text{ is Cohen–Macaulay}\}$$
$$\{\mathfrak{p} \in \mathrm{Spec}(A) \mid A_{\mathfrak{p}} \text{ is Gorenstein}\}$$

are both open in $\mathrm{Spec}(A)$.

Proof. Let $\mathfrak{q} \in X = \mathrm{Spec}(A)$ be such that $A_{\mathfrak{q}}$ is Cohen–Macaulay. Choose a dualizing complex $D^{\cdot}$ normalized such that $H^0(D_{\mathfrak{q}}^{\cdot}) \neq 0$. It follows that $H^i(D_{\mathfrak{q}}^{\cdot}) = 0$ for $i \neq 0$. i.e., that $\mathfrak{q}$ does not belong to the closed set $Z = \bigcup_{i \neq 0} \mathrm{Supp}(H^i(D^{\cdot}))$. Clearly $A_{\mathfrak{p}}$ is Cohen–Macaulay for all $\mathfrak{p} \in X - Z$.

Now suppose in addition $A_{\mathfrak{q}}$ is Gorenstein. Put $D = H^0(D_{\mathfrak{q}}^{\cdot})$. Then with the notation above

$$\{\mathfrak{p} \in X \mid A_{\mathfrak{p}} \text{ is Gorenstein }\} = \{\mathfrak{p} \in X - Z \mid D_{\mathfrak{p}} \simeq A_{\mathfrak{p}}\}.$$

The last set is easily seen to be open. $\qquad \square$

Remark 8.19. The first part of 8.18 is easily seen to generalize as follows. Suppose $M \neq 0$ is a finitely generated A-module. Then

$$\{\mathfrak{p} \in \mathrm{Spec}\, A \mid M_{\mathfrak{p}} \text{ is Cohen–Macaulay}\}$$

is an open subset of $\mathrm{Supp}(M)$.

8.4 Dualizing modules

Let A denote a noetherian ring.

Definition 8.20. A finitely generated module D is called a *dualizing module* if D has a (finite) injective resolution which is a dualizing complex for A.

If D is a dualizing module for A and $D^{\cdot}$ is a finite injective resolution of D, then for all prime ideals $\mathfrak{p}$ in A, $D_{\mathfrak{p}}^{\cdot}$ is a normalized dualizing complex for $A_{\mathfrak{p}}$, as it follows from section 7.5. It follows from 8.16 that $A_{\mathfrak{p}}$ is a Cohen–Macaulay local ring.

Let us remark that a dualizing module is unique in the sense that if $\mathrm{Spec}(A)$ is connected, then any other dualizing module is of the form $D \otimes_A L$, where L is a projective A-module of rank one, as it follows from 7.45.

Proposition 8.21. *The ring A has a dualizing module if and only if it has a dualizing complex and for all prime ideals $\mathfrak{p}$ of A, $A_{\mathfrak{p}}$ is a Cohen–Macaulay ring (**locally Cohen–Macaulay**).*

Proof. Suppose $D^{\cdot}$ is a dualizing complex for A and that A is locally Cohen–Macaulay. We can decompose $A = A_1 \times \cdots \times A_n$ where $\mathrm{Spec}(A_i)$ is connected. Correspondingly we get a decomposition $D^{\cdot} = D_1^{\cdot} \times \cdots \times D_n^{\cdot}$ where $D_i^{\cdot}$ is a dualizing complex for A_i. It will now suffice to prove that each A_i has a dualizing module D_i, $D_1 \times \cdots \times D_n$ will be a dualizing complex for A. From this discussion follows that we may assume that $X = \mathrm{Spec}(A)$ is connected. Put $X_i = \mathrm{Supp}(H^i(D^{\cdot}))$. It follows from the locally Cohen–Macaulayness of A that the X_i's form a partition of X into closed sets. Since only finitely many of these are nonempty, this is a partition into open sets, thus $X = X_j$ for some j. Consequently $H^i(D^{\cdot}) = 0$, $i \neq j$ and whence $H^j(D^{\cdot})$ is a dualizing module for A. $\square$

The following proposition summarizes the basic properties of dualizing modules over a local ring. These are all immediate consequences of our general theory.

Proposition 8.22. *Let A be a noetherian local ring with maximal ideal $\mathfrak{m}$ and residue field k, and let D be a dualizing module for A. Put $d = \dim(A)$.*

(1) *D is a Cohen–Macaulay module with $\mathrm{Supp}(D) = \mathrm{Spec}(A)$.*

(2) *D has a finite injective resolution, such a resolution is a normalized dualizing complex.*

(3) $\operatorname{Tor}_i^A(k, D) \simeq \operatorname{Ext}_A^{d+i}(k, A)$, $i \in \mathbb{Z}$.

(4) $D^\wedge = H_{\mathfrak{m}}^d(A)^\vee$ *where* $-^\wedge$ *denotes the completion and* $-^\vee$ *denotes the Matlis dual.*

Proof. We shall list the principal references separately. (1) 8.16. (2) 7.5. (3) 8.17(3). (4) 8.4 and 6.4. □

Remark 8.23. With the notation of 7.27 let us remark that $D \simeq A$ if and only if A is Gorenstein as it follows from (3) above and uniqueness of the dualizing module. We have a better result.

Corollary 8.24. *If a dualizing module D can be generated by one element, then A is Gorenstein.*

Proof. If D is generated by one element, then $D \otimes_A k \simeq k$ and whence by 8.22(3) we have

$$\operatorname{Ext}_A^d(k, A) \simeq k$$

and consequently A is Gorenstein by 5.35. □

The following proposition is most important for constructing a dualizing module and for calculation of the rank of $\operatorname{Ext}_A^d(k, A)$.

Proposition 8.25. *Let B be a Gorenstein local ring and A a quotient ring of B which admits a resolution by finitely generated free B-modules*

$$0 \longrightarrow L_n \longrightarrow L_{n-1} \longrightarrow \cdots \longrightarrow L_1 \longrightarrow L_0 \longrightarrow A \longrightarrow 0$$

where $n = \dim B - \dim A$. Then A is a Cohen–Macaulay ring with dualizing module $D = \operatorname{Ext}_B^n(A, B)$.

Moreover if $d = \dim A$, the rank of the vector space $\operatorname{Ext}_A^d(k, A)$ is the same as that of the kernel of $k \otimes_A L_n \to k \otimes_A L_{n-1}$.

Proof. The first part follows from the proof of 2.49. Let $I^\bullet$ be a minimal injective resolution of B. $I^\bullet$ is a dualizing complex for B, and $\operatorname{Hom}_B^\bullet(A, I^\bullet)$ is a dualizing complex for A, 7.25. The complex

$$\operatorname{Hom}_B^\bullet(A, I^\bullet)[n]$$

is a normalized dualizing complex for A. A normalized dualizing complex is unique up to homotopy, thus

$$D \simeq H^0(\operatorname{Hom}_B^\bullet(A, I^\bullet)[n]) \simeq \operatorname{Ext}_B^n(A, B)$$

and consequently $D \otimes_A k$ is isomorphic to the cokernel of

$$\mathrm{Hom}_B(L_{n-1}, k) \to \mathrm{Hom}_B(L_n, k).$$

Conclusion by 8.22(3). $\qquad \square$

Example 8.26. If $\mathfrak{J}$ is the ideal generated by $(n-1)$-minors in an $n \times n$-matrix with coefficients in a Gorenstein ring B such that $\mathrm{depth}_{\mathfrak{J}} B = 4$, then $B/\mathfrak{J}$ is Gorenstein, as it follows by considering the Gulliksen–Negaard complex 2.65.

8.5 Locally factorial domains

Let A be a noetherian normal domain with fraction field K. With the notation in Chapter 3 we have the following.

Lemma 8.27. *Let $T \neq 0$ be a finitely generated torsion free module of rank one, $(T \otimes_A K \simeq K)$. Assume for all $\mathfrak{p} \in \mathrm{Spec}(A)$ with $\dim A_{\mathfrak{p}} \geq 2$ that $\mathrm{Ext}^1_{A_{\mathfrak{p}}}(k_{\mathfrak{p}}, T_{\mathfrak{p}}) = 0$. If the first Chern class $c_1(T) \in \mathrm{Cl}(A)$ is zero, then $T \simeq A$.*

Proof. This is contained in the proof of 3.28. $\qquad \square$

Proposition 8.28. *Suppose the noetherian normal domain A has a dualizing module D. Then first Chern class $c_1(D) \in \mathrm{Cl}(A)$ is zero if and only if $D \simeq A$.*

Proof. For any prime ideal $\mathfrak{p}$ of A we have

$$\mathrm{Ext}^{\bullet}_{A_{\mathfrak{p}}}(k_{\mathfrak{p}}, D_{\mathfrak{p}}) \simeq k_{\mathfrak{p}}[-\dim A_{\mathfrak{p}}]$$

as it follows from 7.5. In particular $D \otimes_A K \simeq K$ and for any prime ideal $\mathfrak{p}$ with $\dim A_{\mathfrak{p}} \geq 2$ we have $\mathrm{Ext}^1_{A_{\mathfrak{p}}}(k_{\mathfrak{p}}, D_{\mathfrak{p}}) = 0$.
Conclusion by 8.27. $\qquad \square$

Corollary 8.29. *Suppose for any prime ideal $\mathfrak{p}$ that $A_{\mathfrak{p}}$ is factorial (**locally factorial**) then any dualizing module is projective of rank one.*

Proof. Let us first remark that for any finitely generated module M and any projective module of rank one L we have with the notation of 3.8 that

$$c_1(M \otimes_A L) = c_1(M) + \mathrm{rank}_K(M \otimes_A K)c_1(L).$$

This follows easily from the proof of 3.28. In particular for a dualizing module D we have

$$c_1(D \otimes_A L) = c_1(D) + c_1(L).$$

Since we have an isomorphism 3.29

$$c_1 : \operatorname{Pic}(A) \simeq \operatorname{Cl}(A),$$

we can find a projective module L of rank one such that $c_1(D \otimes_A L) = 0$ and whence by 8.30, $D \otimes_A L \simeq A$. $\qquad\square$

Corollary 8.30. *If A is a locally factorial domain which has a dualizing module, then the module A has finite injective dimension, i.e., A is Gorenstein.*

Proof. Follows from 8.24. $\qquad\square$

8.6   Conductors

In this section we consider an injection $A \to B$ of noetherian rings such that any nonzero divisor in A becomes a nonzero divisor in B and there exists a nonzero divisor in A which annihilates B/A.

From these assumptions follow that B as an A-module is isomorphic to an ideal in A. Thus B is finitely generated as an A-module. Let K denote the ring of fractions of A with respect to the set of all nonzero divisors in A. We leave it to the reader to show that $A \to K$ extends uniquely to a morphism $B \to K$ and that this morphism is an injection. Thus to give an extension as above is the same as to give a subring B of K containing A and being finitely generated as an A-module.

Definition 8.31. The *conductor* $\mathfrak{F}$ of A in B is given by

$$\mathfrak{F} = \operatorname{Ann}_A(B/A).$$

Thus $\mathfrak{F}$ is a nonzero ideal in A. Note that $\mathfrak{F}$ is also an ideal in B, in fact $\mathfrak{F}$ is the largest ideal in B which is contained in A.

Lemma 8.32. *The canonical B-linear map*

$$\mathfrak{F} \to \operatorname{Hom}_A(B, A)$$

is an isomorphism.

Proof. Note that $\mathfrak{F} \to \mathrm{Hom}_A(B, A)$ is an injection since its composite with "evaluation at 1" : $\mathrm{Hom}_A(B, A) \to A$ is the inclusion of $\mathfrak{F}$ in A. Given $f : B \to A$, let the extension of f to an endomorphism of K be scalar multiplication with $c \in K$. We have $c = f(1) \in A$ and $cB \subseteq A$, whence $c \in \mathfrak{F}$. The last part is obvious. $\qquad\square$

Theorem 8.33. *Suppose A is a Gorenstein local ring and B is locally Cohen–Macaulay. Then*

(1) *The conductor $\mathfrak{F}$ is a dualizing module for B.*

(2) *B/A is a dualizing module for $A/\mathfrak{F}$.*

(3) *If $A \neq B$ then $\dim(A/\mathfrak{F}) = \dim(A) - 1$.*

Proof. Let $I^{\bullet}$ be a finite injective resolution of A. By 7.29

$$D^{\bullet} = \mathrm{Hom}_A^{\bullet}(B, I^{\bullet})$$

is a dualizing complex for B. By 8.32 above $\mathfrak{F} \simeq H^0(D^{\bullet})$. By assumption the B-ideal $\mathfrak{F}$ contains a nonzero divisor, thus

$$\mathrm{Supp}(H^0(D^{\bullet})) = \mathrm{Spec}(B).$$

For any prime ideal $\mathfrak{q}$ of B, $D_{\mathfrak{q}}^{\bullet}$ is a dualizing complex for $B_{\mathfrak{q}}$, in particular it has only one nonzero cohomology module since $B_{\mathfrak{q}}$ is Cohen–Macaulay. Thus $H^i(D_{\mathfrak{q}}^{\bullet}) = 0$ for $i \neq 0$. It follows that $H^i(D^{\bullet}) = 0$ for $i \neq 0$, that is

$$\mathrm{Ext}_A^i(B, A) = \begin{cases} 0, & i \neq 0 \\ \mathfrak{F}, & i = 0. \end{cases}$$

This proves that $\mathfrak{F}$ is a dualizing module for B.

We shall now proceed to apply the duality theorem 8.16 for Cohen–Macaulay modules. Consider the exact sequence

$$0 \longrightarrow A \longrightarrow B \longrightarrow B/A \longrightarrow 0.$$

From this we deduce that

$$\mathrm{Ext}_A^i(B/A, A) = 0, \qquad \text{for } i \geq 2$$

and an exact sequence

$$0 \longrightarrow \mathrm{Hom}_A(B/A, A) \longrightarrow \mathrm{Hom}_A(B, A)$$
$$\longrightarrow A \longrightarrow \mathrm{Ext}_A^1(B/A, A) \longrightarrow 0.$$

The map $\mathrm{Hom}_A(B, A) \to A$ is the "evaluation at 1" and therefore an injection with image $\mathfrak{F}$, see the proof of 8.32. In conclusion

$$\mathrm{Ext}_A^i(B/A, A) = \begin{cases} 0, & i \neq 1 \\ A/\mathfrak{F}, & i = 1. \end{cases}$$

This proves that B/A is a Cohen–Macaulay module of dimension one less that the dimension of A. The module dual to B/A is $A/\mathfrak{F}$, thus $A/\mathfrak{F}$ is Cohen–Macaulay of dimension $\dim(A) - 1$ and

$$\mathrm{Ext}_A^1(A/\mathfrak{F}, A) = \begin{cases} 0 & \text{for } i \neq 1 \\ B/A & \text{for } i = 1. \end{cases}$$

By 7.25 the complex $\mathrm{Hom}_A^{\bullet}(A/\mathfrak{F}, I^{\bullet})$ is a dualizing complex for $A/\mathfrak{F}$, whence $\mathrm{Ext}_A^1(A/\mathfrak{F}, A) = B/A$ is a dualizing module for $A/\mathfrak{F}$. $\qquad \square$

Corollary 8.34. *If A is a Gorenstein local ring of dimension one then B is locally Cohen–Macaulay, $A/\mathfrak{F}$ is an Artin ring and we have equality of length*

$$\ell_A(B/A) = \ell_A(A/\mathfrak{F}).$$

Proof. Let us first remark that the A-module B is Cohen–Macaulay. Any nonzero divisor in A is a nonzero divisor for B, thus $\mathrm{depth}_A B \neq 0$. Since A is one dimensional this implies that B is Cohen–Macaulay A-module. The first part of the proof of 8.33 shows in fact that this is equivalent to B being locally Cohen–Macaulay.

The conductor $\mathfrak{F}$ contains a nonzero divisor, whence $A/\mathfrak{F}$ is an Artin ring. For any Artin ring, its dualizing module and the ring itself are modules of the same length as it follows from Matlis duality. $\qquad \square$

Corollary 8.35. *If A is a Gorenstein local ring and B is locally Gorenstein, then the conductor $\mathfrak{F}$ considered as an ideal in B is generated by one nonzero divisor.*

Proof. Let us first prove that B contains only finitely many ideals. Indeed for any maximal ideal $\mathfrak{n}$ of B we have $\mathfrak{n} \cap A = \mathfrak{m}$, the maximal ideal of A. Namely, given $x \neq 0$, $x \in A/\mathfrak{n} \cap A$, x is invertible in $B/\mathfrak{n}$, thus x^{-1} satisfies an integral equation

$$x^{-n} + a_1 x^{-n+1} + \cdots + a_n = 0, \qquad a_i \in A/\mathfrak{n} \cap A$$

from which we deduce

$$-x^{-1} = a_1 + \cdots + a_n x^{n-1}$$

which show $x^{-1} \in A/\mathfrak{n} \cap A$. The ring $B/\mathfrak{m}B$ is artinian admitting only finitely many maximal ideals in B.

By 8.33 and 8.23 we have for each prime ideal $\mathfrak{q}$ in B, $\mathfrak{F}_{\mathfrak{q}} \simeq B_{\mathfrak{q}}$. It follows that $\mathfrak{F}$ is a projective module of rank one. If $\mathfrak{n}_1, \ldots, \mathfrak{n}_s$ are the maximal ideals in B, then $\mathfrak{F}/\mathfrak{n}_1 \cdots \mathfrak{n}_s\mathfrak{F}$ is a free $B/\mathfrak{n}_1 \cdots \mathfrak{n}_s$-module of rank one. By Nakayama's lemma a basis element lifts to a basis element of $\mathfrak{F}$.   □

Example 8.36 (Ordinary surface singularities). Let k denote a field of characteristic not 2.

(1) *double point*

$$k[[X, Y, Z]]/(XY) \to k[[S, U]] \times k[[T, U]]$$
$$(X, Y, Z) \mapsto ((S, 0), (0, T), (U, U))$$
$$\mathfrak{F} = ((S, T)).$$

(2) *pinch point*

$$k[[X, Y, Z]]/(X^2 - YZ^2) \to k[[S, T]]$$
$$(X, Y, Z) \mapsto (ST, S^2, T)$$
$$\mathfrak{F} = (T).$$

(3) *triple point*

$$k[[X, Y, Z]]/(XYZ) \to k[[S, T]] \times k[[T, U]] \times k[[S, U]]$$
$$(X, Y, Z) \mapsto ((S, T, 0), (0, T, U), (S, 0, U))$$
$$\mathfrak{F} = ((S, T, U)).$$

8.7 Formal fibers

A ring homorphism $f : A \to B$ of local rings is called a *local homomorphism* if the maximal ideal in A is mapped into the maximal ideal in B.

Let $X^{\cdot}$ be a bounded below complex with finitely generated cohomology modules. For a prime ideal $\mathfrak{p}$ we define the Poincaré series

$$\mu_{\mathfrak{p}}(X^{\cdot}, t) = \sum_i \mathrm{rank}_{k_{\mathfrak{p}}}(\mathrm{Ext}^i_{A_{\mathfrak{p}}}(k_{\mathfrak{p}}, X^{\cdot}_{\mathfrak{p}}))t^i = \mu(X^{\cdot}_{\mathfrak{p}}, t).$$

Proposition 8.37. *Let $f : A \to B$ be a flat local homomorphism of noethe-rian local rings, $\mathfrak{m}$ and $\mathfrak{n}$ denote the maximal ideals in A and B. For a bounded below complex $X^{\cdot}$ of A-modules with finitely generated cohomology we have*

$$\mu_{\mathfrak{m}}(X^{\cdot} \otimes_A B, t) = \mu_{\mathfrak{m}}(X^{\cdot}, t)\mu_{\mathfrak{n}}(B/\mathfrak{m}B, t),$$

where the last term is relative to the $B/\mathfrak{m}B$-module $B/\mathfrak{m}B$.

Proof. Choose a quasi-isomorphism $X^{\cdot} \to D^{\cdot}$ into a bounded below complex of injective A-modules, and a quasi-isomorphism $D^{\cdot} \otimes_A B \to E^{\cdot}$ into a bounded below complex of injective B-modules. Note that this defines a morphism

$$\mathrm{Hom}_A^{\cdot}(X^{\cdot}, D^{\cdot}) \otimes_A B \to \mathrm{Hom}_B^{\cdot}(X^{\cdot} \otimes_A B, E^{\cdot}).$$

Note that this map is a quasi-isomorphism in case $X^{\cdot} = A$. By the method developed in 7.14, this is seen to be a quasi-isomorphism whenever $X^{\cdot}$ is a bounded above complex with finitely generated cohomology. In particular for the residue field k of A we have a quasi-isomorphism

$$\mathrm{Hom}_A^{\cdot}(k, D^{\cdot}) \otimes_A B \to \mathrm{Hom}_B^{\cdot}(k \otimes_A B, E^{\cdot}).$$

Let now k' denote the residue field of $k \otimes_A B$, and let $P^{\cdot}$ be a resolution of k' by finitely generated free $k \otimes_A B$-modules. We deduce a quasi-isomorphism, 7.12,

$$\mathrm{Hom}_{k \otimes_A B}^{\cdot}(P^{\cdot}, \mathrm{Hom}_B^{\cdot}(k \otimes_A B, E^{\cdot})) \to \mathrm{Hom}_{k \otimes_A B}^{\cdot}(P^{\cdot}, \mathrm{Hom}_A^{\cdot}(k, D^{\cdot}) \otimes_A B).$$

The left hand side is isomorphic to (compare the proof of 7.25)

$$\mathrm{Hom}_B^{\cdot}(P^{\cdot}, E^{\cdot}) \simeq \mathrm{Hom}_B^{\cdot}(k', E^{\cdot}).$$

The right hand side is isomorphic to

$$\mathrm{Hom}_{k \otimes_A B}^{\cdot}(P^{\cdot}, k \otimes_A B) \otimes_{k \otimes_A B} \mathrm{Hom}_A^{\cdot}(k, D^{\cdot}) \otimes_A B$$
$$\simeq \mathrm{Hom}_{k \otimes_A B}^{\cdot}(P^{\cdot}, k \otimes_A B) \otimes_k \mathrm{Hom}_A^{\cdot}(k, D^{\cdot})$$

Conclusion by passing to cohomology.   □

Corollary 8.38. *Let $f : A \to B$ be a flat local homomorphism of noetherian local rings and suppose that $k \otimes_A B$ is a Gorenstein local ring. If A has a dualizing complex $D^{\cdot}$, then B has a dualizing complex $E^{\cdot}$ for which there exists a quasi-isomorphism*

$$D^{\cdot} \otimes_A B \to E^{\cdot}.$$

Moreover for any prime ideal $\mathfrak{p}$ in A, the ring $k_{\mathfrak{p}} \otimes_A B$ is Gorenstein.

Proof. Choose a bounded below complex $E^{\cdot}$ of injective modules and a quasi-isomorphism $D^{\cdot} \otimes_A B \to E^{\cdot}$. By definition $\mu_n(E^{\cdot}, t) = \mu_n(D^{\cdot} \otimes_A B, t)$. By assumption we have $\mu_m(D^{\cdot}, t) = t^r$ and $\mu_n(k \otimes_A B, t) = t^s$ for some $r, s \in \mathbb{Z}$, whence

$$\mu_n(E^{\cdot}, t) = t^{r+s}.$$

By Proposition 7.17 we may assume that $E^{\cdot}$ is bounded. It follows from 7.36 that $E^{\cdot}$ is a dualizing complex for B.

For the second part we may assume that A is an integral domain and that $\mathfrak{p} = 0$. Let K denote the fraction field of A. Let $D^{\cdot}$ and $E^{\cdot}$ be dualizing complexes for A and B, connected by a quasi-isomorphism

$$D^{\cdot} \otimes_A B \to E^{\cdot}.$$

From this we deduce a quasi-isomorphism

$$D^{\cdot} \otimes_A B \otimes_A K \to E^{\cdot} \otimes_B (B \otimes_A K).$$

Note $D^{\cdot} \otimes_A B \otimes_A K \simeq (D^{\cdot} \otimes_A K) \otimes_K (B \otimes_A K)$ and that $D^{\cdot} \otimes_A K$ is a dualizing complex for K. Thus we can find $d \in \mathbb{Z}$ and a quasi-isomorphism

$$K[d] \to D^{\cdot} \otimes_A K$$

and whence a quasi-isomorphism

$$B \otimes_A K[d] \to E^{\cdot} \otimes_B (B \otimes_A K),$$

from which we deduce that $B \otimes_A K$ has a finite injective resolution. $\quad\square$

Remark 8.39. Let $\mathfrak{p}$ be a prime ideal in a noetherian local ring A and let $k_{\mathfrak{p}}$ denote the residue field of $A/\mathfrak{p}$. The ring $k_{\mathfrak{p}} \otimes_A \hat{A}$ is called the *formal fiber* of A at $\mathfrak{p}$. It follows from 7.38 that if A has a dualizing complex, then all formal fibers of A are Gorenstein local rings.

It should be remarked that Ferrand and Raynaud have constructed a noetherian local ring of dimension one, whose formal fiber at $\mathfrak{p} = (0)$ is not a Gorenstein ring.

The above remark can be generalized as follows.

Corollary 8.40. *Let $D^{\cdot}$ be a bounded complex of injective A-modules with finitely generated cohomology. If a prime ideal $\mathfrak{p}$ in A is such that $H^{\cdot}(D^{\cdot}_{\mathfrak{p}}) \neq 0$, then the formal fiber $k_{\mathfrak{p}} \otimes_A \hat{A}$ is Gorenstein.*

Proof. By 8.37 we have

$$\mu_{\hat{\mathfrak{m}}}(D^{\cdot} \otimes_A \hat{A}, t) = \mu_{\mathfrak{m}}(D^{\cdot}, t).$$

Thus we can find a bounded complex of injective $\hat{A}$-modules $E^{\cdot}$ and a quasi-isomorphism

$$D^{\cdot} \otimes_A \hat{A} \to E^{\cdot}.$$

For a prime ideal $\mathfrak{q}$ in $k_{\mathfrak{p}} \otimes_A \hat{A}$ consider the flat local homomorphism $A_{\mathfrak{p}} \to \hat{A}_{\mathfrak{q}}$. We have

$$\mu_{\mathfrak{q}}(D^{\cdot} \otimes_A \hat{A}_{\mathfrak{q}}, t) = \mu_{\mathfrak{p}}(D_{\mathfrak{p}}^{\cdot}, t)\mu_{\mathfrak{q}}(k_{\mathfrak{p}} \otimes_A \hat{A}_{\mathfrak{q}}, t).$$

Using that $\mu_{\mathfrak{p}}(D_{\mathfrak{p}}^{\cdot}, t) \neq 0$ we deduce that $\mu_{\mathfrak{q}}(D^{\cdot} \otimes_A \hat{A}_{\mathfrak{q}}, t)$ is a polynomial with

$$\deg \mu_{\mathfrak{q}}(E^{\cdot} \otimes_A \hat{A}_{\mathfrak{q}}, t) \leq \deg \mu_{\mathfrak{q}}(E_{\mathfrak{q}}^{\cdot}, t).$$

Recall that $E^{\cdot}$ is a bounded complex of injective modules and deduce that

$$\mu_{\mathfrak{q}}(D^{\cdot} \otimes_A \hat{A}_{\mathfrak{q}}, t)$$

is bounded as $\mathfrak{q}$ varies through all prime ideals in $k_{\mathfrak{p}} \otimes_A \hat{A}$. Conclusion by 7.15. □

Corollary 8.41. *Let A be a noetherian ring. If A has a dualizing complex $D^{\cdot}$, then $A[X]$ has a dualizing complex $E^{\cdot}$ for which there exists a quasi-isomorphism*

$$D^{\cdot} \otimes_A A[X] \to E^{\cdot}.$$

Proof. Let $\mathfrak{q}$ be a prime ideal in $A[X]$ and set $\mathfrak{p} = \mathfrak{q} \cap A$. We have

$$\mu_{\mathfrak{q}}(D^{\cdot} \otimes_A A[X], t) = \mu_{\mathfrak{p}}(D_{\mathfrak{p}}^{\cdot}, t)\mu_{\mathfrak{q}}(k_{\mathfrak{p}}[X], t).$$

Clearly $\mu_{\mathfrak{q}}(k_{\mathfrak{p}}[X], t) = t^r$, $r = 0, 1$. Conclusion by 7.15 and 7.36. □

Chapter 9

Amplitude and Dimension

Let A be a ring. By a module is understood an A-module and by a complex is understood a complex of A-modules.

Given a complex $E^{\bullet}$ of injective modules and a module M then we put

$$\text{Ext}_A^{\bullet}(M, E^{\bullet}) = \text{Hom}_A^{\bullet}(M, E^{\bullet}).$$

For a bounded below complex $X^{\bullet}$ we choose a quasi-isomorphism $X^{\bullet} \to E^{\bullet}$ into a bounded below complex of injective modules, 7.6, and define

$$\text{Ext}_A^{\bullet}(M, X^{\bullet}) = \text{Ext}_A^{\bullet}(M, E^{\bullet}).$$

By 7.4 this is independent of choice.

Given a complex $P^{\bullet}$ of projective modules and a module M then we put

$$\text{Tor}_{\bullet}^A(M, P^{\bullet}) = M \otimes_A P^{\bullet}.$$

For a bounded above complex $X^{\bullet}$ we choose a quasi-isomorphism $P^{\bullet} \to X^{\bullet}$ from a bounded above complex of projective modules, 7.40, and define

$$\text{Tor}_{\bullet}^A(M, X^{\bullet}) = \text{Tor}_{\bullet}^A(M, P^{\bullet}).$$

By 7.12 this is independent of choice.

Note the index convention

$$\text{Tor}_i^A(M, X^{\bullet}) = H^{-i}(\text{Tor}_{\bullet}^A(M, X^{\bullet})).$$

In this rest of this chapter A denotes a noetherian local ring with maximal ideal $\mathfrak{m}$ and residue field k.

9.1   Depth of a complex

For a complex $X^\bullet$ with finitely generated cohomology modules $H^\bullet(X^\bullet) \neq 0$ we make the following conventions.

If $X^\bullet$ is bounded above we put

$$s(X^\bullet) = \sup\{i \mid H^i(X^\bullet) \neq 0\}$$

and if $X^\bullet$ is bounded below

$$i(X^\bullet) = \inf\{i \mid H^i(X^\bullet) \neq 0\}.$$

In case $X^\bullet$ is bounded we define the *amplitude* of $X^\bullet$

$$\mathrm{amp}(X^\bullet) = s(X^\bullet) - i(X^\bullet)$$

and also the *length*

$$\ell(X^\bullet) = \sup\{i \mid X^i \neq 0\} - \inf\{i \mid X^i \neq 0\}.$$

Definition 9.1. Given a complex $X^\bullet$, $H^\bullet(X^\bullet) \neq 0$. If $X^\bullet$ is bounded we define the *depth*

$$\mathrm{depth}(X^\bullet) = i(\mathrm{Ext}^\bullet_A(k, X^\bullet)) - s(\mathrm{Tor}^A_\bullet(k, X^\bullet)).$$

If $X^\bullet$ has finite projective amplitude we define *projective amplitude*

$$\mathrm{proj.\,amp}(X^\bullet) = s(\mathrm{Tor}^A_\bullet(k, X^\bullet)) - i(\mathrm{Tor}^A_\bullet(k, X^\bullet))$$

and if $X^\bullet$ has finite injective amplitude, *injective amplitude*

$$\mathrm{inj.\,amp}(X^\bullet) = s(\mathrm{Ext}^\bullet_A(k, X^\bullet)) - i(\mathrm{Ext}^\bullet_A(k, X^\bullet)).$$

By 8.2 this gives well defined integers. Note also by Nakayama's lemma

$$s(\mathrm{Tor}^A_\bullet(k, X^\bullet)) = s(X^\bullet).$$

For $X^\bullet$ a module, the depth is the usual depth and the projective amplitude is the projective dimension, but the injective amplitude is usually not the injective dimension.

Remark 9.2. These invariants can be calculated from the Poincaré series, recall 8.1,

$$\mu(X^{\boldsymbol{\cdot}}, t) = \sum_i \operatorname{rank}_k(\operatorname{Ext}_A^i(k, X^{\boldsymbol{\cdot}}))t^i,$$

$$\beta(X^{\boldsymbol{\cdot}}, t) = \sum_i \operatorname{rank}_k(\operatorname{Tor}_i^A(k, X^{\boldsymbol{\cdot}}))t^i.$$

For a Laurent series $f(t) = \sum a_i t^i$ we have the signed zero order

$$\nu_t(f(t)) = \inf\{i \mid a_i \neq 0\}.$$

Note the effect of the index convention

$$\nu_t(\beta(X^{\boldsymbol{\cdot}}, t)) = -s(\operatorname{Tor}_{\boldsymbol{\cdot}}^A(k, X^{\boldsymbol{\cdot}})).$$

Then we have

$$\operatorname{depth}(X^{\boldsymbol{\cdot}}) = \nu_t(\beta(X^{\boldsymbol{\cdot}}, t)\mu(X^{\boldsymbol{\cdot}}, t)),$$
$$\operatorname{proj.amp}(X^{\boldsymbol{\cdot}}) = -\nu_t(\beta(X^{\boldsymbol{\cdot}}, t)\beta(X^{\boldsymbol{\cdot}}, t^{-1})),$$
$$\operatorname{inj.amp}(X^{\boldsymbol{\cdot}}) = -\nu_t(\mu(X^{\boldsymbol{\cdot}}, t)\mu(X^{\boldsymbol{\cdot}}, t^{-1})).$$

Remark 9.3. If the cohomology modules of $X^{\boldsymbol{\cdot}}$ have finite length, we have $i(\operatorname{Ext}_A^{\boldsymbol{\cdot}}(k, X^{\boldsymbol{\cdot}})) = i(X^{\boldsymbol{\cdot}})$ giving

$$\operatorname{depth}(X^{\boldsymbol{\cdot}}) = -\operatorname{amp}(X^{\boldsymbol{\cdot}}).$$

Proposition 9.4. *Let $L^{\boldsymbol{\cdot}}$ be a bounded complex of finitely generated free modules with $H^{\boldsymbol{\cdot}}(L^{\boldsymbol{\cdot}}) \neq 0$ and $X^{\boldsymbol{\cdot}}$ a bounded complex as above, then*

$$\operatorname{depth}(X^{\boldsymbol{\cdot}} \otimes_A L^{\boldsymbol{\cdot}}) = \operatorname{depth}(X^{\boldsymbol{\cdot}}) - \operatorname{proj.amp}(L^{\boldsymbol{\cdot}}).$$

Proof. From 8.8 we get

$$\mu(X^{\boldsymbol{\cdot}} \otimes_A L^{\boldsymbol{\cdot}}, t) = \mu(X^{\boldsymbol{\cdot}}, t)\beta(L^{\boldsymbol{\cdot}}, t^{-1})$$

clearly

$$\beta(X^{\boldsymbol{\cdot}} \otimes_A L^{\boldsymbol{\cdot}}, t) = \beta(X^{\boldsymbol{\cdot}}, t)\beta(L^{\boldsymbol{\cdot}}, t).$$

Multiply these two together, apply ν_t and use the formulas in 9.2. □

The following corollary generalizes 2.47 and 2.60.

Corollary 9.5. *With the notation of 9.4 we have*

$$\operatorname{depth}(L^{\textstyle\cdot}) + \operatorname{proj.} \operatorname{amp}(L^{\textstyle\cdot}) = \operatorname{depth}(A)$$

and

$$\operatorname{depth}(X^{\textstyle\cdot} \otimes_A L^{\textstyle\cdot}) + \operatorname{depth}(A) = \operatorname{depth}(X^{\textstyle\cdot}) + \operatorname{depth}(L^{\textstyle\cdot}).$$

If the cohomology modules of $X^{\textstyle\cdot} \otimes_A L^{\textstyle\cdot}$ have finite length, then

$$\operatorname{amp}(X^{\textstyle\cdot} \otimes_A L^{\textstyle\cdot}) = \operatorname{proj.} \operatorname{amp}(L^{\textstyle\cdot}) - \operatorname{depth}(X^{\textstyle\cdot}).$$

Proof. We have

$$\beta(\operatorname{Hom}_A^{\textstyle\cdot}(L^{\textstyle\cdot}, A), t) = \beta(L^{\textstyle\cdot}, t^{-1})$$

and by 7.3

$$\mu(L^{\textstyle\cdot}, t) = \beta(L^{\textstyle\cdot}, t)\mu(A, t).$$

Conclude by applying ν_t. $\qquad\qquad\square$

In the rest of this section $E^{\textstyle\cdot}$ denotes a bounded complex of injective modules with finitely generated cohomology modules and $H^{\textstyle\cdot}(E^{\textstyle\cdot}) \neq 0$.

Proposition 9.6. *For $X^{\textstyle\cdot}$ a bounded complex with finitely generated cohomology modules and $H^{\textstyle\cdot}(X^{\textstyle\cdot}) \neq 0$.*

$$\operatorname{depth}(\operatorname{Hom}_A^{\textstyle\cdot}(X^{\textstyle\cdot}, E^{\textstyle\cdot})) = \operatorname{depth}(X^{\textstyle\cdot}) - \operatorname{inj.} \operatorname{amp}(E^{\textstyle\cdot}).$$

If $X^{\textstyle\cdot}$ has finite projective amplitude

$$\operatorname{inj.} \operatorname{amp}(\operatorname{Hom}_A^{\textstyle\cdot}(X^{\textstyle\cdot}, E^{\textstyle\cdot})) = \operatorname{proj.} \operatorname{amp}(X^{\textstyle\cdot}) + \operatorname{inj.} \operatorname{amp}(E^{\textstyle\cdot}).$$

If $X^{\textstyle\cdot}$ has finite injective amplitude

$$\operatorname{proj.} \operatorname{amp}(\operatorname{Hom}_A^{\textstyle\cdot}(X^{\textstyle\cdot}, E^{\textstyle\cdot})) = \operatorname{inj.} \operatorname{amp}(X^{\textstyle\cdot}) + \operatorname{inj.} \operatorname{amp}(E^{\textstyle\cdot}).$$

Proof. According to 8.3 we have

$$\mu(\operatorname{Hom}_A^{\textstyle\cdot}(X^{\textstyle\cdot}, E^{\textstyle\cdot}), t) = \beta(X^{\textstyle\cdot}, t)\mu(E^{\textstyle\cdot}, t),$$
$$\beta(\operatorname{Hom}_A^{\textstyle\cdot}(X^{\textstyle\cdot}, E^{\textstyle\cdot}), t) = \mu(X^{\textstyle\cdot}, t)\mu(E^{\textstyle\cdot}, t^{-1}),$$

from which the result is easily deduced from the formulas in 9.2. $\qquad\square$

Remark 9.7. The formulas in 9.6 are particularly interesting when $E^{\textstyle\cdot}$ is a dualizing complex. In this case $\mu(E^{\textstyle\cdot}, t) = t^d$ for some $d \in \mathbb{Z}$, thus

$$\operatorname{inj.} \operatorname{amp}(E^{\textstyle\cdot}) = 0.$$

Remark 9.8. From 9.6 follows in particular that

$$\operatorname{depth}(E^{\cdot}) + \operatorname{inj. amp}(E^{\cdot}) = \operatorname{depth}(A).$$

In case $E^{\cdot}$ is minimal it follows from 7.15 that $s(E^{\cdot}) = s(\operatorname{Ext}_A^{\cdot}(k, E^{\cdot}))$, so the length of $E^{\cdot}$ equals

$$\begin{aligned} \ell(E^{\cdot}) &= \operatorname{amp}(E^{\cdot}) + \operatorname{depth}(E^{\cdot}) + \operatorname{inj. amp}(E^{\cdot}) \\ &= \operatorname{amp}(E^{\cdot}) + \operatorname{depth}(A). \end{aligned}$$

In particular, if $E^{\cdot}$ is a minimal resolution of a finitely generated module M of finite injective dimension, we get

$$\operatorname{inj dim}(M) = \operatorname{depth}(A).$$

Lemma 9.9. *For any finitely generated module $M \neq 0$ we have*

$$\operatorname{depth}(M) \leq \ell(E^{\cdot}).$$

Proof. From the formula

$$\beta(\operatorname{Hom}_A^{\cdot}(M, E^{\cdot}), t) = \mu(M, t)\mu(E^{\cdot}, t^{-1})$$

we get

$$-s(\operatorname{Hom}_A^{\cdot}(M, E^{\cdot})) = \operatorname{depth}(M) - s(\operatorname{Ext}_A^{\cdot}(k, E^{\cdot}))$$

and further

$$s(\operatorname{Ext}_A^{\cdot}(k, E^{\cdot})) - i(E^{\cdot}) = \operatorname{depth}(M) + s(\operatorname{Hom}_A^{\cdot}(M, E^{\cdot})) - i(E^{\cdot}).$$

The term $s(\operatorname{Hom}_A^{\cdot}(M, E^{\cdot})) - i(E^{\cdot})$ is positive, so

$$s(\operatorname{Ext}_A^{\cdot}(k, E^{\cdot})) - i(E^{\cdot}) \geq \operatorname{depth}(M)$$

but clearly

$$\ell(E^{\cdot}) \geq s(\operatorname{Ext}_A^{\cdot}(k, E^{\cdot})) - i(E^{\cdot}). \qquad \square$$

Proposition 9.10. *Let $M \neq 0$ be a finitely generated module. Let M have Poincaré series*

$$\mu(M, t) = \sum \mu_i(M)t^i.$$

Then the set of $i \in \mathbb{N}$, for which $\mu_i(M) \neq 0$ is an interval.

Proof. It suffices to prove the proposition under the presence of a dualizing complex $D^{\cdot}$ (pass to the completion of A). We shall assume $D^{\cdot}$ normalized such that

$$\mu(D^{\cdot}, t) = 1.$$

For the complex $\operatorname{Hom}_A^{\cdot}(M, D^{\cdot})$ we have

$$\beta(\operatorname{Hom}_A^{\cdot}(M, D^{\cdot}), t) = \mu(M, t).$$

Chose a quasi-isomorphism $L^{\cdot} \to \operatorname{Hom}_A^{\cdot}(M, D^{\cdot})$ where $L^{\cdot}$ is a bounded above complex of finitely generated free modules. We have

$$\beta(L^{\cdot}, t) = \mu(M, t).$$

Let us now assume that $\operatorname{Ext}_A^s(k, M) = 0$ where $s > \operatorname{depth}(M)$. This gives $\operatorname{Tor}_s^A(k, L^{\cdot}) = 0$. Let us now look at the complex $L^{\cdot}$

$$L_{s+1} \xrightarrow{\ \partial_{s+1}\ } L_s \xrightarrow{\ \partial_s\ } L_{s-1}.$$

We have the exact sequence

$$0 \longrightarrow \operatorname{Tor}_s^A(k, L^{\cdot}) \longrightarrow k \otimes_A \operatorname{Cok}(\partial_{s+1}) \longrightarrow k \otimes_A L_{s+1}.$$

It follows from 2.34 that

$$\operatorname{Cok}(\partial_{s+1}) \to L_{s+1}$$

has a retraction. Consider now the two complexes

$$L_+^{\cdot} : 0 \longrightarrow \operatorname{Cok}(\partial_s) \longrightarrow L_{s-2} \longrightarrow L_{s-3} \longrightarrow \cdots$$
$$L_-^{\cdot} : \cdots \longrightarrow L_{s+2} \longrightarrow L_{s+1} \longrightarrow \operatorname{Ker}(\partial_s) \longrightarrow 0.$$

We have

$$L^{\cdot} \simeq L_+^{\cdot} \oplus L_-^{\cdot}$$

with

$$\beta(L_+^{\cdot}, t) = \cdots + \mu_{s-2}(M)t^{s-2} + \mu_{s-1}(M)t^{s-1}$$
$$\beta(L_-^{\cdot}, t) = \mu_{s+1}(M)t^{s+1} + \mu_{s+2}(M)t^{s+2} + \cdots.$$

By duality we have a quasi-isomorphism of complexes

$$M \simeq \operatorname{Hom}_A^{\cdot}(L_+^{\cdot}, D^{\cdot}) \oplus \operatorname{Hom}_A^{\cdot}(L_-^{\cdot}, D^{\cdot}).$$

In particular the two complexes $\text{Hom}^{\bullet}_A(L^{\bullet}_+, D^{\bullet})$ and $\text{Hom}^{\bullet}_A(L^{\bullet}_-, D^{\bullet})$ have only cohomology in level zero. Put

$$M_+ = H^0(\text{Hom}^{\bullet}_A(L^{\bullet}_+, D^{\bullet})),$$
$$M_- = H^0(\text{Hom}^{\bullet}_A(L^{\bullet}_-, D^{\bullet})).$$

In recapitulation

$$M \simeq M_+ \oplus M_-$$

with

$$\mu(M_+, t) = \cdots + \mu_{s-2}(M)t^{s-2} + \mu_{s-1}(M)t^{s-1}$$
$$\mu(M_-, t) = \mu_{s+1}(M)t^{s+1} + \mu_{s+2}(M)t^{s+2} + \cdots.$$

Since $s > \text{depth}\, M$, we have $\mu(M_+, t) \neq 0$ and whence $M_+ \neq 0$. If M_- were different from zero, we could have $\text{depth}(M_-) > s$, contradicting 9.9 when applied to M_- and a minimal injective resolution of M_+. Thus

$$\mu(M, t) = \cdots + \mu_{s-2}(M)t^{s-2} + \mu_{s-1}(M)t^{s-1}. \qquad \square$$

Let us close this section with another application of the methods developed here.

Proposition 9.11. *Let $P \neq 0$ be a finitely generated module of projective dimension n. Then for any finitely generated module $M \neq 0$*

$$\text{Ext}^n_A(P, M) \neq 0.$$

Proof. Let $L^{\bullet}$ be a free resolution of P. By 8.3 $\mu(\text{Hom}^{\bullet}_A(L^{\bullet}, M), t) = \beta(L^{\bullet}, t)\mu(M, t)$ from which we conclude

$$i(\text{Ext}^n_A(k, \text{Hom}^{\bullet}_A(L^{\bullet}, M))) = \text{depth}(M)$$

and whence

$$\text{depth}(\text{Hom}^{\bullet}_A(L^{\bullet}, M)) = \text{depth}(M) - s(\text{Hom}^{\bullet}_A(L^{\bullet}, M)).$$

On the other hand, by 9.4

$$\text{depth}(\text{Hom}^{\bullet}_A(L^{\bullet}, A) \otimes_A M) = \text{depth}(M) - \text{proj.\,amp}(\text{Hom}^{\bullet}_A(L^{\bullet}, A))$$

but $\text{Hom}^{\bullet}_A(L^{\bullet}, A) \otimes_A M \simeq \text{Hom}^{\bullet}_A(L^{\bullet}, M)$ and $\text{proj.\,amp}(\text{Hom}^{\bullet}_A(L^{\bullet}, A)) = n$, so

$$s(\text{Hom}^{\bullet}_A(L^{\bullet}, M)) = n. \qquad \square$$

9.2   The dual of a module

In this section we let d denote the dimension of our local ring A, and let $D^{\bullet}$ be a normalized dualizing complex.

Proposition 9.12. *Let $N \neq 0$ be a finitely generated module of dimension n. Then*

(1) $\dim(\mathrm{Ext}^{i}_{A}(N, D^{\bullet})) \leq d - i$.

(2) $\mathrm{Ext}^{i}_{A}(N, D^{\bullet}) = 0$ *for* $i < d - n$.

(3) $\mathrm{Ass}(\mathrm{Ext}^{d-n}_{A}(N, D^{\bullet})) = \{\mathfrak{p} \in \mathrm{Supp}(N) \mid \dim(A/\mathfrak{p}) = n\}$.

Proof. We shall assume that $D^{\bullet}$ is minimal.

To prove (1) we need to estimate

$$\sup\{\dim(A/\mathfrak{p}) \mid \mathfrak{p} \in \mathrm{Supp}(\mathrm{Ext}^{i}_{A}(N, D^{\bullet}))\}$$

so let $\mathfrak{p} \in \mathrm{Supp}(\mathrm{Ext}^{i}_{A}(N, D^{\bullet}))$. This means $\mathrm{Ext}^{i}_{A}(N_{\mathfrak{p}}, D^{\bullet}_{\mathfrak{p}}) \neq 0$. By minimality of $D^{\bullet}_{\mathfrak{p}}$ we have $D^{j}_{\mathfrak{p}} = 0$ for $j > d - \dim(A/\mathfrak{p})$, whence $i \leq d - \dim(A/\mathfrak{p})$ or $\dim(A/\mathfrak{p}) \leq d - i$. This proves (1).

To prove (2) let us first remark that 2.15 and 7.35 give

$$\begin{aligned} \mathrm{Ass}(\mathrm{Hom}_{A}(N, D^{i})) &= \mathrm{Supp}(N) \cap \mathrm{Ass}(D^{i}) \\ &= \{\mathfrak{p} \in \mathrm{Supp}(N) \mid \dim(A/\mathfrak{p}) = d - i\}. \end{aligned}$$

This shows that the complex $\mathrm{Hom}^{\bullet}_{A}(N, D^{\bullet})$ has the form

$$0 \longrightarrow \mathrm{Hom}_{A}(N, D^{d-n}) \longrightarrow \mathrm{Hom}_{A}(N, D^{d-n+1}) \longrightarrow \cdots$$

proving (2) and one half of (3),

$$\mathrm{Ass}(\mathrm{Ext}^{d-n}_{A}(N, D^{\bullet})) \subseteq \{\mathfrak{p} \in \mathrm{Supp}(N) \mid \dim(A/\mathfrak{p}) = n\}.$$

Conversely, let $\mathfrak{p} \in \mathrm{Supp}(N)$ with $\dim(A/\mathfrak{p}) = n$. By assertion (2) we have $\mathrm{Ext}^{j}_{A}(N_{\mathfrak{p}}, D^{\bullet}_{\mathfrak{p}}) = 0$ for $j < d - n$, and from (1) applied to $A_{\mathfrak{p}}$

$$\dim(\mathrm{Ext}^{j}_{A}(N_{\mathfrak{p}}, D^{\bullet}_{\mathfrak{p}})) \leq (d - n) - j.$$

This shows that $\mathrm{Ext}^{j}_{A}(N_{\mathfrak{p}}, D^{\bullet}_{\mathfrak{p}}) = 0$ for $j > d - n$ and that $\mathrm{Ext}^{d-n}_{A}(N, D^{\bullet})$ has finite length. Since $\mathrm{Ext}^{j}_{A}(N, D^{\bullet})_{\mathfrak{p}} = 0$ for $j \neq d - n$ and $\mathfrak{p} \in \mathrm{Supp}(N)$ we must have $\mathrm{Ext}^{d-n}_{A}(N, D^{\bullet})_{\mathfrak{p}} \neq 0$ and since this module has finite length

$$\mathfrak{p} \in \mathrm{Ass}(\mathrm{Ext}^{d-n}_{A}(N, D^{\bullet})). \qquad \square$$

Theorem 9.13. *Let $N \neq 0$ be a finitely generated module, then*

$$\dim(N) = \mathrm{depth}(N) + \mathrm{amp}(\mathrm{Hom}_A^{\bullet}(N, D^{\bullet})).$$

Proof. By 9.6 and 9.7 we have $\mathrm{depth}(N) = \mathrm{depth}(\mathrm{Hom}_A^{\bullet}(N, D^{\bullet}))$ we shall therefore calculate

$$\mathrm{depth}(\mathrm{Hom}_A^{\bullet}(N, D^{\bullet})) + \mathrm{amp}(\mathrm{Hom}_A^{\bullet}(N, D^{\bullet}))$$
$$= i(\mathrm{Ext}_A^{\bullet}(k, \mathrm{Hom}_A^{\bullet}(N, D^{\bullet})) - i(\mathrm{Hom}_A^{\bullet}(N, D^{\bullet})).$$

The first term we can calculate from 9.14 below

$$i(\mathrm{Ext}_A^{\bullet}(k, \mathrm{Hom}_A^{\bullet}(N, D^{\bullet})) = i(\mathrm{Ext}_A^{\bullet}(k, D^{\bullet})) = \dim(A).$$

The second term we can read off 9.12

$$i(\mathrm{Hom}_A^{\bullet}(N, D^{\bullet})) = \dim(A) - \dim(N). \qquad \square$$

Lemma 9.14. *Let $X^{\bullet}$ be a bounded above complex and $E^{\bullet}$ a bounded below complex of injective modules. Suppose both have finitely generated cohomology modules and $H^{\bullet}(X^{\bullet}) \neq 0$ and $H^{\bullet}(E^{\bullet}) \neq 0$. Then*

$$i(\mathrm{Ext}_A^{\bullet}(k, \mathrm{Hom}_A^{\bullet}(X^{\bullet}, E^{\bullet})) = i(\mathrm{Ext}_A^{\bullet}(k, E^{\bullet})) - s(X^{\bullet}).$$

Proof. By 8.3 we have

$$\mu(\mathrm{Hom}_A^{\bullet}(X^{\bullet}, E^{\bullet}), t) = \beta(X^{\bullet}, t)\mu(E^{\bullet}, t).$$

Conclusion using ν_t. $\qquad \square$

9.3   The amplitude formula

Let $D^{\bullet}$ be a dualizing complex for the local ring A. The purpose of this section is to generalize formula 9.13

$$\dim(N) = \mathrm{depth}(N) + \mathrm{amp}(\mathrm{Hom}_A^{\bullet}(N, D^{\bullet}))$$

to complexes. Thus we will set out to calculate

$$\mathrm{depth}(X^{\bullet}) + \mathrm{amp}(\mathrm{Hom}_A^{\bullet}(N, D^{\bullet})),$$

where $X^{\bullet}$ is a bounded complex with finitely generated cohomology modules and $H^{\bullet}(X^{\bullet}) \neq 0$. The starting point is 9.16 below applied to $\mathrm{Hom}_A^{\bullet}(X^{\bullet}, D^{\bullet})$.

Definition 9.15. Let $Z^{\bullet}$ be a complex with finitely generated cohomology modules. Define the *support* of $Z^{\bullet}$ by

$$\mathrm{Supp}(Z^{\bullet}) = \bigcup_i \mathrm{Supp}(H^i(Z^{\bullet})).$$

Proposition 9.16. *Let $Z^{\bullet}$ be a bounded below complex with finitely generated cohomology modules and suppose that $H^{\bullet}(Z^{\bullet})) \neq 0$. Then*

$$i(Z^{\bullet}) = \inf\{i(\mathrm{Ext}_A^{\bullet}(k_{\mathfrak{p}}, Z_{\mathfrak{p}}^{\bullet}) \mid \mathfrak{p} \in \mathrm{Supp}(Z^{\bullet})\}.$$

Proof. We may assume that $Z^i = 0$ for $i < 0$ and $H^0(Z^{\bullet}) \neq 0$. Now clearly the equality above has a positive right hand term. Let us remark that for any prime ideal $\mathfrak{p}$ we have

$$\mathrm{Hom}_A(k_{\mathfrak{p}}, H^0(Z_{\mathfrak{p}}^{\bullet})) = \mathrm{Ext}_A^0(k_{\mathfrak{p}}, Z_{\mathfrak{p}}^{\bullet}).$$

Now consider a $\mathfrak{p} \in \mathrm{Ass}(H^0(Z^{\bullet}))$ to finish the proof. $\square$

Theorem 9.17. *Let $X^{\bullet}$ be a bounded complex with finitely generated cohomology modules and suppose that $H^{\bullet}(X^{\bullet})) \neq 0$. Then*

$$\mathrm{depth}(X^{\bullet}) + \mathrm{amp}(\mathrm{Hom}_A^{\bullet}(X^{\bullet}, D^{\bullet}))$$
$$= \sup\{\dim(H^j(X^{\bullet})) + j - s(X^{\bullet}) \mid H^j(X^{\bullet}) \neq 0\}.$$

Proof. Since $\mathrm{depth}(X^{\bullet}) = \mathrm{depth}(\mathrm{Hom}_A^{\bullet}(X^{\bullet}, D^{\bullet}))$ start to calculate

$$\mathrm{depth}(\mathrm{Hom}_A^{\bullet}(X^{\bullet}, D^{\bullet})) + \mathrm{amp}(\mathrm{Hom}_A^{\bullet}(X^{\bullet}, D^{\bullet}))$$
$$= i(\mathrm{Ext}_A^{\bullet}(k, \mathrm{Hom}_A^{\bullet}(X^{\bullet}, D^{\bullet}))) - i(\mathrm{Hom}_A^{\bullet}(X^{\bullet}, D^{\bullet})).$$

At this point it is convenient to let $D^{\bullet}$ be a normalized dualizing complex. To calculate the first term, we use 9.14

$$i(\mathrm{Ext}_A^{\bullet}(k, \mathrm{Hom}_A^{\bullet}(X^{\bullet}, E^{\bullet}))) = \dim(A) - s(X^{\bullet}).$$

Let us at once record a localized form of this. For $\mathfrak{p} \in \mathrm{Supp}(X^{\bullet})$ we have

$$i(\mathrm{Ext}_A^{\bullet}(k_{\mathfrak{p}}, \mathrm{Hom}_A^{\bullet}(X_{\mathfrak{p}}^{\bullet}, E_{\mathfrak{p}}^{\bullet}))) = \dim(A) - \dim(A/\mathfrak{p}) - s(X_{\mathfrak{p}}^{\bullet}).$$

Using this and the previous lemma applied to $Z^{\bullet} = \mathrm{Hom}_A^{\bullet}(X^{\bullet}, D^{\bullet})$ we get

$$i(\mathrm{Hom}_A^{\bullet}(X^{\bullet}, D^{\bullet})) = \inf\{\dim(A) - \dim(A/\mathfrak{p}) - s(X_{\mathfrak{p}}^{\bullet}) \mid \mathfrak{p} \in \mathrm{Supp}(X^{\bullet})\}.$$

Collect this together to get

$$\text{depth}(\text{Hom}_A^{\bullet}(X^{\bullet}, D^{\bullet})) + \text{amp}(\text{Hom}_A^{\bullet}(X^{\bullet}, D^{\bullet}))$$
$$= -s(X^{\bullet}) + \sup\{\dim(A/\mathfrak{p}) + s(X_{\mathfrak{p}}^{\bullet}) \mid \mathfrak{p} \in \text{Supp}(X^{\bullet})\}.$$

The evolution of this term will be the subject of a separate lemma. □

Lemma 9.18. *Let* $X^{\bullet}$ *be a bounded complex with finitely generated cohomology modules and suppose that* $H^{\bullet}(X^{\bullet})) \neq 0$. *Then*

$$\sup\{\dim(A/\mathfrak{p}) + s(X_{\mathfrak{p}}^{\bullet}) \mid \mathfrak{p} \in \text{Supp}(X^{\bullet})\}$$
$$= \sup\{\dim(H^j(X^{\bullet})) + j \mid H^j(X^{\bullet}) \neq 0\}.$$

Proof. We give the elementary calculation

$$\sup\{\dim(A/\mathfrak{p}) + s(X_{\mathfrak{p}}^{\bullet}) \mid \mathfrak{p} \in \text{Supp}(X^{\bullet})\}$$
$$= \sup\{\dim(A/\mathfrak{p}) + j \mid H^j(X_{\mathfrak{p}}^{\bullet}) \neq 0\}$$
$$= \sup\{\dim(A/\mathfrak{p}) + j \mid \mathfrak{p} \in \text{Supp}(H^j(X^{\bullet}))\}$$
$$= \sup\{\dim(H^j(X^{\bullet})) + j \mid H^j(X^{\bullet}) \neq 0\}.$$

□

9.4 Dimension of a complex

Let A be a noetherian local ring with maximal ideal $\mathfrak{m}$ and residue field k. For a bounded complex $X^{\bullet}$ with finitely generated cohomology modules and $H^{\bullet}(X^{\bullet}) \neq 0$ we make the following definition.

Definition 9.19. The *dimension* of the complex $X^{\bullet}$ is

$$\dim(X^{\bullet}) = \sup\{\dim(H^j(X^{\bullet})) + j - s(X^{\bullet}) \mid H^j(X^{\bullet}) \neq 0\}.$$

Remark 9.20. If the complex $X.$ has the form

$$0 \longrightarrow X_n \longrightarrow X_{n-1} \longrightarrow \cdots \longrightarrow X_1 \longrightarrow X_0 \longrightarrow 0$$

with $H_0(X.) \neq 0$, then we have

$$\dim(X.) = \sup\{\dim(H_j(X.)) - j \mid H_j(X.) \neq 0\}.$$

Remark 9.21. It is clear from the definition, that

$$0 \leq \dim(X^{\bullet}) \leq \dim(\text{Supp}(X^{\bullet})).$$

Proposition 9.22. *Let $M \neq 0$ be a finitely generated module and $L.$ a bounded complex of finitely generated free modules*

$$0 \longrightarrow L_s \longrightarrow L_{s-1} \longrightarrow \cdots \longrightarrow L_1 \longrightarrow L_0 \longrightarrow 0.$$

If $H_0(L.) \neq 0$, then

$$\dim(M \otimes_A L.) = \sup\{\dim(M \otimes_A H_j(L.)) - j \mid H_j(M \otimes_A L.) \neq 0\}.$$

Proof. By 9.18 we have

$$\dim(M \otimes_A L.) = \sup\{\dim(A/\mathfrak{p}) + s(M_\mathfrak{p} \otimes_A L._\mathfrak{p}) \mid \mathfrak{p} \in \mathrm{Supp}(M \otimes_A L.)\}$$

so given $\mathfrak{p} \in \mathrm{Supp}(M \otimes_A L.)$, let $j \in \mathbb{N}$ be such that $s(M_\mathfrak{p} \otimes_A L._\mathfrak{p}) = -j$. Since $L.$ is a complex of free modules we have $s(M_\mathfrak{p} \otimes_A L._\mathfrak{p}) = s(L._\mathfrak{p})$, whence $\mathfrak{p} \in \mathrm{Supp}(M \otimes_A L.)$ and

$$\dim(A/\mathfrak{p}) + s(M_\mathfrak{p} \otimes_A L._\mathfrak{p}) \leq \dim(M \otimes_A H_j(L.)) - j.$$

Conversely, given $j \in \mathbb{N}$ such that $H_j(L.) \neq 0$ choose $\mathfrak{p} \in \mathrm{Supp}(M \otimes_A H_j(L.))$ such that $\dim(A/\mathfrak{p}) = \dim(M \otimes_A H_j(L.)) - j$ we have $-j \leq s(M_\mathfrak{p} \otimes_{A_\mathfrak{p}} L._\mathfrak{p})$ and whence

$$\dim(A/\mathfrak{p}) + s(M_\mathfrak{p} \otimes_{A_\mathfrak{p}} L._\mathfrak{p}) \geq \dim(M \otimes_A H_j(L.)) - j. \qquad \square$$

We shall now restate the fundamental duality between dimension and amplitude.

Theorem 9.23. *Suppose A has a dualizing complex $D^{\cdot}$. Then for a complex $X^{\cdot}$ as above*

$$\dim(X^{\cdot}) = \mathrm{depth}(X^{\cdot}) + \mathrm{amp}(\mathrm{Hom}_A^{\cdot}(X^{\cdot}, D^{\cdot})).$$

Proof. This is a reformulation of 9.17. $\qquad \square$

We shall take the opportunity to improve the inequality 9.21. Let us remark that $\mathrm{Supp}(X^{\cdot}) = \mathrm{Supp}(\mathrm{Hom}_A^{\cdot}(X^{\cdot}, D^{\cdot}))$.

Proposition 9.24. *For the complex $X^{\cdot}$ put $c = i(\mathrm{Hom}_A^{\cdot}(X^{\cdot}, D^{\cdot}))$. We have*

$$\dim(X^{\cdot}) \leq \inf\{\dim(A/\mathfrak{p}) \mid \mathfrak{p} \in \mathrm{Ass}(H^c(\mathrm{Hom}_A^{\cdot}(X^{\cdot}, D^{\cdot})))\}.$$

Moreover, $\dim(X^{\cdot}) = 0$ if and only if $\mathfrak{m} \in \mathrm{Ass}(H^c(\mathrm{Hom}_A^{\cdot}(X^{\cdot}, D^{\cdot})))\}$.

Proof. Suppose $\mathfrak{p} \in \mathrm{Ass}(H^c(\mathrm{Hom}_A^\bullet(X^\bullet, D^\bullet)))$. Since

$$\mathrm{Hom}_A(k_\mathfrak{p}, H^c(\mathrm{Hom}_A^\bullet(X_\mathfrak{p}^\bullet, D_\mathfrak{p}^\bullet))) \simeq \mathrm{Ext}_A^c(k_\mathfrak{p}, \mathrm{Hom}_A^\bullet(X_\mathfrak{p}^\bullet, D_\mathfrak{p}^\bullet))$$

we have by 7.16

$$\mathrm{Ext}_A^{c+d}(k_\mathfrak{p}, H^c(\mathrm{Hom}_A^\bullet(X_\mathfrak{p}^\bullet, D_\mathfrak{p}^\bullet))) \neq 0; \qquad d = \dim(A/\mathfrak{p}).$$

From this follows

$$\mathrm{depth}(\mathrm{Hom}_A^\bullet(X^\bullet, D^\bullet)) + \mathrm{amp}(\mathrm{Hom}_A^\bullet(X^\bullet, D^\bullet))$$
$$= i(\mathrm{Ext}_A^\bullet(k, \mathrm{Hom}_A^\bullet(X^\bullet, D^\bullet))) - i(\mathrm{Hom}_A^\bullet(X^\bullet, D^\bullet)) \leq c + d - c = d,$$

proving the first part. To prove the second part, $\dim(X^\bullet) = 0$ if and only if $\mathrm{Ext}_A^c(k, \mathrm{Hom}_A^\bullet(X^\bullet, D^\bullet)) \neq 0$. As noticed above this is the same as $\mathrm{Hom}_A^c(k, H^c(\mathrm{Hom}_A^\bullet(X^\bullet, D^\bullet))) \neq 0$. $\qquad\square$

The reader will notice that the result of this chapter has been established without the use of local cohomology. We will now apply our results to local cohomology, compare section 8.2.

Theorem 9.25. *For a bounded complex $X^\bullet$ as above*

$$s(R^\bullet\Gamma_\mathfrak{m}(X^\bullet)) = \sup\{\dim(H^j(X^\bullet)) + j \mid H^j(X^\bullet) \neq 0\}$$

otherwise expressed

$$\dim(X^\bullet) = s(R^\bullet\Gamma_\mathfrak{m}(X^\bullet)) - s(X^\bullet).$$

Proof. Passing to the completion we may assume that A has a dualizing $D^\bullet$. By 9.26 below we have

$$i(R^\bullet\Gamma_\mathfrak{m}(X^\bullet)) = i(\mathrm{Ext}_A^\bullet(k, X^\bullet)).$$

By the local duality theorem 7.17

$$\mathrm{amp}(R^\bullet\Gamma_\mathfrak{m}(X^\bullet)) = \mathrm{amp}(\mathrm{Hom}_A^\bullet(X^\bullet, D^\bullet))$$

combining these two we get

$$s(R^\bullet\Gamma_\mathfrak{m}(X^\bullet)) = i(\mathrm{Ext}_A^\bullet(k, X^\bullet)) + \mathrm{amp}\,\mathrm{Hom}_A^\bullet(X^\bullet, D^\bullet)$$

or by subtracting $s(X^\bullet)$ on both sides

$$s(R^\bullet\Gamma_\mathfrak{m}(X^\bullet)) - s(X^\bullet) = \mathrm{depth}(X^\bullet) + \mathrm{amp}(\mathrm{Hom}_A^\bullet(X^\bullet, D^\bullet))$$

conclusion by 9.23. $\qquad\square$

Lemma 9.26. *Let $X^{\cdot}$ be a bounded below complex with finitely generated cohomology modules and $H^{\cdot}(X^{\cdot}) \neq 0$. Then*

$$i(R^{\cdot}\Gamma_{\mathfrak{m}}(X^{\cdot})) = i(\operatorname{Ext}_A^{\cdot}(k, X^{\cdot})).$$

Proof. Choose a quasi-isomorphism $X^{\cdot} \to E^{\cdot}$ into a minimal injective complex. We have

$$\operatorname{Hom}_A^{\cdot}(k, E^{\cdot}) = \operatorname{Hom}_A^{\cdot}(k, \Gamma_{\mathfrak{m}}(E^{\cdot})).$$

Notice that the complex $\Gamma_{\mathfrak{m}}(E^{\cdot})$ is again minimal. Therefore if $c = \inf\{i \mid \Gamma_{\mathfrak{m}}(E^i)\}$, then $\Gamma_{\mathfrak{m}}(E^c)$ is an essential extension of

$$\operatorname{Ker}(\Gamma_{\mathfrak{m}}(E^c) \to \Gamma_{\mathfrak{m}}(E^{c+1}))$$

thus this kernel is nontrivial and therefore $c = i(\Gamma_{\mathfrak{m}}(E^{\cdot}))$. We conclude by the exact sequence

$$0 \longrightarrow \operatorname{Hom}_A(k, H^c(\Gamma_{\mathfrak{m}}(E^{\cdot}))) \longrightarrow H^c(\operatorname{Hom}_A(k, E^{\cdot})). \qquad \square$$

9.5   The tensor product formula

In this section we shall derive an important formula for calculating dimension of a tensor product of complexes. We consider a bounded complex $L_{\cdot}$ of finitely generated free modules with $H_{\cdot}(L_{\cdot}) \neq 0$ and a complex $X_{\cdot}$ with finitely generated cohomology modules

$$0 \longrightarrow X_n \longrightarrow X_{n-1} \longrightarrow \cdots \longrightarrow X_1 \longrightarrow X_0 \longrightarrow 0$$

with $H_0(X_{\cdot}) \neq 0$.

Theorem 9.27 (Tensor product formula).

$$\dim(X_{\cdot} \otimes_A L_{\cdot}) = \sup\{\dim(H_j(X_{\cdot}) \otimes_A L_{\cdot}) - j \mid H_j(X_{\cdot}) \neq 0\}.$$

Proof. For notational convenience we will assume $L_i = 0$ for $i < 0$ and $H_0(L_{\cdot}) \neq 0$. Now using 9.18 we calculate the dimension.

$$\dim(X_{\cdot} \otimes_A L_{\cdot})$$
$$= \sup\{\dim(H_j(X_{\cdot} \otimes_A L_{\cdot})) - j \mid H_j(X_{\cdot} \otimes_A L_{\cdot}) \neq 0\}$$
$$= \sup\{\dim(A/\mathfrak{p}) + s(X_{\cdot\mathfrak{p}} \otimes_{A_{\mathfrak{p}}} L_{\cdot\mathfrak{p}}) \mid \mathfrak{p} \in \operatorname{Supp}(X_{\cdot}) \cap \operatorname{Supp}(L_{\cdot})\}.$$

We have the elementary formula

$$s(X._{\mathfrak{p}} \otimes_{A_{\mathfrak{p}}} L._{\mathfrak{p}}) = \sup\{s(H_j(X._{\mathfrak{p}}) \otimes_{A_{\mathfrak{p}}} L._{\mathfrak{p}})) - j \mid H_j(X._{\mathfrak{p}}) \neq 0\}.$$

Thus we get the following expression

$$\dim(X. \otimes_A L.)$$
$$= \sup\{\dim(A/\mathfrak{p}) + s(H_j(X._{\mathfrak{p}}) \otimes_A L._{\mathfrak{p}}) - j \mid \mathfrak{p} \in \mathrm{Supp}(H_j(X.) \otimes_A L.)\}$$

now using 9.18 once more we get

$$\dim(X. \otimes_A L.) = \sup\{\dim(H_j(X.) \otimes_A L.) - j \mid H_j(X.) \neq 0\}. \qquad \square$$

We shall here reexamine the principle expressed in the tensor product formula.

Lemma 9.28. *Let $E^{\cdot}$ be a bounded below complex of injective modules and $X^{\cdot}$ a bounded complex. Suppose both complexes have finitely generated cohomology modules and $H^{\cdot}(X^{\cdot}) \neq 0$, $H^{\cdot}(E^{\cdot}) \neq 0$. Then*

$$i(\mathrm{Hom}_A^{\cdot}(X^{\cdot}, E^{\cdot})) = \inf\{i(\mathrm{Hom}_A^{\cdot}(H^j(X^{\cdot}), E^{\cdot})) - j \mid H^j(X^{\cdot}) \neq 0\}.$$

Proof. By 9.16 and 9.14

$$i(\mathrm{Hom}_A^{\cdot}(X^{\cdot}, E^{\cdot}))$$
$$= \inf\{i(\mathrm{Ext}_A^{\cdot}(k_{\mathfrak{p}}, \mathrm{Hom}_A^{\cdot}(X^{\cdot}, E_{\mathfrak{p}}^{\cdot}))) \mid H^{\cdot}(X_{\mathfrak{p}}^{\cdot}) \neq 0\}$$
$$= \inf\{i(\mathrm{Ext}_A^{\cdot}(k_{\mathfrak{p}}, E_{\mathfrak{p}}^{\cdot})) - s(X_{\mathfrak{p}}^{\cdot}) \mid H^{\cdot}(X_{\mathfrak{p}}^{\cdot}) \neq 0\}$$
$$= \inf\{i(\mathrm{Ext}_A^{\cdot}(k_{\mathfrak{p}}, E_{\mathfrak{p}}^{\cdot})) + j \mid \mathfrak{p} \in \mathrm{Supp}(E^{\cdot}), H^j(X_{\mathfrak{p}}^{\cdot}) \neq 0\}.$$

From this we get by substituting $H^j(X^{\cdot})$ for $X^{\cdot}$

$$i(\mathrm{Hom}_A^{\cdot}(H^j(X^{\cdot}), E^{\cdot}))$$
$$= \inf\{i(\mathrm{Ext}_A^{\cdot}(k_{\mathfrak{p}}, E_{\mathfrak{p}}^{\cdot})) \mid \mathfrak{p} \in \mathrm{Supp}(E^{\cdot}) \cap \mathrm{Supp}(H^j(X^{\cdot}))\}.$$

Thus we get

$$i(\mathrm{Hom}_A^{\cdot}(X^{\cdot}, E^{\cdot})) = \inf\{i(\mathrm{Hom}_A^{\cdot}(H^j(X^{\cdot}), E^{\cdot})) - j \mid H^j(X^{\cdot}) \neq 0\}. \qquad \square$$

9.6   Depth inequalities

Let A denote a noetherian ring and $\mathfrak{I}$ an ideal in A. Throughout this section $X^{\bullet}$ denotes a complex with finitely generated cohomology modules and $H^{\bullet}(X^{\bullet}) \neq 0$.

Definition 9.29.

$$\mathrm{depth}_{\mathfrak{I}}(X^{\bullet}) = i(\mathrm{Ext}_A^{\bullet}(A/\mathfrak{I}, X^{\bullet})) - s(X^{\bullet}).$$

Note that if $\mathrm{Ext}_A^{\bullet}(A/\mathfrak{I}, X^{\bullet}) = 0$, then $\mathrm{depth}_{\mathfrak{I}}(X^{\bullet}) = +\infty$.

Proposition 9.30. *Let N be a finitely generated module with support,* $\mathrm{Supp}(N) = V(\mathfrak{I})$, *then*

$$\mathrm{depth}_{\mathfrak{I}}(X^{\bullet}) = i(\mathrm{Ext}_A^{\bullet}(N, X^{\bullet})) - s(X^{\bullet}).$$

Proof. By 9.14 and 9.16 we have

$$i(\mathrm{Ext}_A^{\bullet}(N, X^{\bullet})) = \inf\{i(\mathrm{Ext}_A^{\bullet}(k_{\mathfrak{p}}, X_{\mathfrak{p}}^{\bullet})) \mid \mathfrak{p} \in \mathrm{Supp}(N) \cap \mathrm{Supp}(X^{\bullet})\}. \qquad \square$$

Corollary 9.31. *Let $\mathfrak{I}$ and $\mathfrak{J}$ be two ideals in A. Then*

$$\mathrm{depth}_{\mathfrak{I}\mathfrak{J}}(X^{\bullet}) = \inf\{\mathrm{depth}_{\mathfrak{I}}(X^{\bullet}), \mathrm{depth}_{\mathfrak{J}}(X^{\bullet})\}.$$

Proof. We have
$$\mathrm{Supp}(A/\mathfrak{I}\mathfrak{J}) = \mathrm{Supp}(A/\mathfrak{I} \oplus A/\mathfrak{J}). \qquad \square$$

Proposition 9.32. *If $\mathrm{Supp}(X^{\bullet}) \subseteq V(\mathfrak{I})$, then*

$$\mathrm{depth}_{\,} I(X^{\bullet}) = -\mathrm{amp}(X^{\bullet}).$$

Proof. Let us assume that $X^i = 0$ for $i < 0$ and $H^0(X^{\bullet}) \neq 0$. We have

$$\mathrm{Ext}_A^0(A/\mathfrak{I}, X^{\bullet}) = \mathrm{Hom}_A(A/\mathfrak{I}, H^0(X^{\bullet})).$$

On the other hand, by 2.15

$$\mathrm{Ass}(\mathrm{Hom}_A(A/\mathfrak{I}, H^0(X^{\bullet})) = V(\mathfrak{I}) \cap \mathrm{Ass}(H^0(X^{\bullet}))$$

and

$$\mathrm{Ass}(H^0(X^{\bullet})) \subseteq \mathrm{Supp}(X^{\bullet}) \subseteq V(\mathfrak{I})$$

so

$$\mathrm{Ass}(\mathrm{Hom}_A(A/\mathfrak{I}, H^0(X^{\bullet})) = \mathrm{Ass}(H^0(X^{\bullet})). \qquad \square$$

The following is known as the *acyclicity lemma*.

Lemma 9.33. *Consider a complex*

$$0 \longrightarrow E_s \longrightarrow E_{s-1} \longrightarrow \cdots \longrightarrow E_0 \longrightarrow 0$$

with finitely generated cohomology modules, such that for $r = 1, \ldots, s$

(1) $\operatorname{Supp}(H_r(E.)) \subseteq V(\mathfrak{I})$.

(2) $\operatorname{Ext}_A^i(A/\mathfrak{I}, E_r) = 0$ *for* $i < r$.

Then

$$H_s(E.) = H_{s-1}(E.) = \cdots = H_1(E.) = 0.$$

Proof. Condition (2) alone implies

$$i(\operatorname{Ext}_A^{\cdot}(A/\mathfrak{I}, E.)) = 0 \geq 0$$

as one sees by an easy induction on s. Let E' denote the complex

$$0 \longrightarrow E_s \longrightarrow E_{s-1} \longrightarrow \cdots \longrightarrow E_1 \longrightarrow \operatorname{Im}(\partial_0) \longrightarrow 0.$$

Let us assume $H.(E'.) \neq 0$. Of course as above $i(\operatorname{Ext}_A^{\cdot}(A/\mathfrak{I}, E'.)) = 0 \geq 0$. Whence

$$\operatorname{depth}_{\mathfrak{I}}(E'.) \geq 0.$$

But, by 9.32 $\operatorname{depth}_{\mathfrak{I}}(E'.) = -\operatorname{amp}(E'.) \leq 0$ and therefore $\operatorname{depth}_{\mathfrak{I}}(E'.) = 0$. This implies $s(E'.) = 0$, a contradiction. $\qquad\square$

Theorem 9.34. *Let* $X^{\cdot}$ *be a bounded complex with finitely generated cohomology modules,* $L^{\cdot}$ *a bounded complex of finitely generated free modules. If* $\operatorname{Supp}(X^{\cdot}) \cap \operatorname{Supp}(L^{\cdot}) \neq \emptyset$, *then*

$$\operatorname{depth}_{\mathfrak{I}}(X^{\cdot} \otimes_A L^{\cdot}) \geq \operatorname{depth}_{\mathfrak{I}}(X^{\cdot}) - \ell(L^{\cdot}).$$

Proof. Put $L^{\cdot\vee} = \operatorname{Hom}_A^{\cdot}(L^{\cdot}, A)$ and choose a quasi-isomorphism $X^{\cdot} \to E^{\cdot}$ where $E^{\cdot}$ is a bounded below complex of injective modules. $X^{\cdot} \otimes_A L^{\cdot} \to E^{\cdot} \otimes_A L^{\cdot}$ is a quasi-isomorphism and the second complex is a complex of injective modules. Thus

$$i(\operatorname{Ext}_A^{\cdot}(A/\mathfrak{I}, X^{\cdot} \otimes_A L^{\cdot})) = i(\operatorname{Hom}_A^{\cdot}(A/\mathfrak{I}, E^{\cdot} \otimes_A L^{\cdot})).$$

We have the following natural isomorphisms

$$\mathrm{Hom}^{\cdot}_A(A/\mathfrak{I}, E^{\cdot} \otimes_A L^{\cdot}) \simeq \mathrm{Hom}^{\cdot}_A(A/\mathfrak{I}, \mathrm{Hom}^{\cdot}_A(L^{\cdot \vee}, E^{\cdot}))$$
$$\simeq \mathrm{Hom}^{\cdot}_A(L^{\cdot \vee} \otimes_A A/\mathfrak{I}, \mathrm{Hom}^{\cdot}_A(A/\mathfrak{I}, E^{\cdot})).$$

Note that $L^{\cdot \vee} \otimes_A A/\mathfrak{I}$ is a complex of projective $A/\mathfrak{I}$-modules and further that $\mathrm{Hom}^{\cdot}_A(A/\mathfrak{I}, E^{\cdot})$ is a complex of injective $A/\mathfrak{I}$-modules. Thus, if one of these complexes has no cohomology it is homotopy equivalent to the zero complex and we get $\mathrm{depth}_{\mathfrak{I}}(X^{\cdot} \otimes_A L^{\cdot}) = +\infty$. Thus we may assume that they both have nonzero cohomology. We get

$$i(\mathrm{Hom}^{\cdot}_A(A/\mathfrak{I}, E^{\cdot} \otimes_A L^{\cdot})) \geq i(\mathrm{Hom}^{\cdot}_A(A/\mathfrak{I}, E^{\cdot})) - s(L^{\cdot \vee} \otimes_A A/\mathfrak{I})$$

or

$$i(\mathrm{Ext}^{\cdot}_A(A/\mathfrak{I}, X^{\cdot} \otimes_A L^{\cdot})) \geq i(\mathrm{Ext}^{\cdot}_A(A/\mathfrak{I}, X^{\cdot})) - s(L^{\cdot \vee} \otimes_A A/\mathfrak{I}).$$

Note $s(X^{\cdot} \otimes_A L^{\cdot}) \leq s(X^{\cdot}) + s(L^{\cdot})$ and $s(L^{\cdot \vee} \otimes_A A/\mathfrak{I}) \leq s(L^{\cdot \vee})$, thus

$$\mathrm{depth}_{\mathfrak{I}}(X^{\cdot} \otimes_A L^{\cdot}) \geq \mathrm{depth}_{\mathfrak{I}}(X^{\cdot}) - (s(L^{\cdot}) + s(L^{\cdot \vee})).$$

We leave it to the reader to prove that

$$\ell(L^{\cdot}) \leq s(L^{\cdot}) + s(L^{\cdot \vee}). \qquad \square$$

We leave it to the reader to recover theorem 2.60 and theorem 2.63 from 9.34.

Proposition 9.35.

$$\mathrm{depth}_{\mathfrak{I}}(X^{\cdot}) = i(R^{\cdot}\Gamma_{\mathfrak{I}}(X^{\cdot})) - s(X^{\cdot}).$$

Proof. The proof of 9.26 generalizes to prove

$$i(R^{\cdot}\Gamma_{\mathfrak{I}}(X^{\cdot})) = i(\mathrm{Ext}^{\cdot}_A(A/\mathfrak{I}, X^{\cdot})). \qquad \square$$

Proposition 9.36. *Let* $f : B \to A$ *be a morphism of noetherian rings such that* A *is finitely generated as* B-module *and* $\mathfrak{I}$ *is an ideal in* B *with* $\mathfrak{I} = Af(\mathfrak{I})$. *Then*

$$\mathrm{depth}_{\mathfrak{I}}(X^{\cdot}) = \mathrm{depth}_{\mathfrak{I}}(X^{\cdot})$$

where on the left hand side we consider $X^{\cdot}$ *as a complex of* B-modules via f.

Proof. It follows from the proof of 6.4 that $R^{\cdot}\Gamma_{\mathfrak{I}}(X^{\cdot}) \simeq R^{\cdot}\Gamma_{\mathfrak{I}}(X^{\cdot})$. $\qquad \square$

Proposition 9.37. *Let us suppose A is local. Given a bounded complex $L.$ of finitely generated free modules,*

$$0 \longrightarrow L_s \longrightarrow L_{s-1} \longrightarrow \cdots \longrightarrow L_0 \longrightarrow 0$$

with $H_0(L.) \neq 0$. For $r = 0, 1, \ldots, s$, let $\mathfrak{I}_r$ denote an ideal such that

$$\mathrm{Supp}(H^r(\mathrm{Hom}_A^{\cdot}(L., A))) = V(\mathfrak{I}_r).$$

For any bounded complex $X.$ with finitely generated cohomology modules and with $H.(X.) \neq 0$, we have

$$\mathrm{amp}(X. \otimes_A L.) = \sup\{r - \mathrm{depth}_{\mathfrak{I}_r}(X.) \mid r = 0, \ldots, s\}.$$

Proof. Choose a quasi-isomorphism $X. \to E.$ where $E.$ is a bounded below complex of injective modules, and apply 9.28 to the complexes $\mathrm{Hom}_A^{\cdot}(L., A)$ and $E..$ □

Using local cohomology 9.35, we can make 9.32 more explicit.

Lemma 9.38. *If $\mathrm{Supp}(X^{\cdot}) \subseteq V(\mathfrak{I})$, then there is a quasi-isomorphism*

$$X^{\cdot} \to R^{\cdot}\Gamma_{\mathfrak{I}}(X^{\cdot}).$$

Proof. Choose a quasi-isomorphism $X^{\cdot} \to E^{\cdot}$ into a minimal injective complex $E^{\cdot}$. Let $\{mod\}$ denote the category of A-modules and $\{mod_{\mathfrak{I}}\}$ the category of A-modules with support in $V(\mathfrak{I})$. The inclusion functor

$$\{mod_{\mathfrak{I}}\} \to \{mod\}$$

and the section functor

$$\Gamma_{\mathfrak{I}} : \{mod_{\mathfrak{I}}\} \to \{mod\}$$

transform injectives into injectives and essential extensions into essential extensions as it follows from section 6.1. From this follows easily that $\Gamma_{\mathfrak{I}}(E^{\cdot}) \simeq E^{\cdot}$. □

As an example of application, let us consider the *Euler-characteristic* of a bounded complex $L^{\cdot}$ whose cohomology modules have finite length, we put

$$\chi(L^{\cdot}) = \sum_i (-1)^i \ell_A(H^i(L^{\cdot})).$$

Proposition 9.39. *Suppose A is a local ring and consider a bounded complex $L^{\cdot}$ of finitely generated free modules whose cohomology modules have finite length. If A is Gorenstein of dimension d, then*

$$\chi(\operatorname{Hom}_A^{\cdot}(L^{\cdot}, A)) = (-1)^d \chi(L^{\cdot}).$$

Proof. By 9.38 we have

$$L^{\cdot} \simeq R^{\cdot}\Gamma_{\mathfrak{m}}(L^{\cdot}).$$

Let $D^{\cdot}$ be a minimal injective resolution of A. The complex $D^{\cdot}$ is a normalized dualizing complex for A. Thus by the local duality theorem 8.14

$$R^{\cdot}\Gamma_{\mathfrak{m}}(L^{\cdot})[d] \simeq \operatorname{Hom}_A^{\cdot}(L^{\cdot}, D^{\cdot})^{\vee}$$

where $-^{\vee}$ denotes the Matlis dual. We have a quasi-isomorphism

$$\operatorname{Hom}_A^{\cdot}(L^{\cdot}, D^{\cdot}) \simeq \operatorname{Hom}_A^{\cdot}(L^{\cdot}, A).$$

Taking these quasi-isomorphisms into account we get that $H^{i+d}(L^{\cdot})$ and $H^i(\operatorname{Hom}_A^{\cdot}(L^{\cdot}, A))$ are Matlis dual to each other, and therefore

$$\ell_A(H^{i+d}(L^{\cdot})) = \ell_A(H^i(\operatorname{Hom}_A^{\cdot}(L^{\cdot}, A))). \qquad \square$$

9.7   Condition S_r of Serre

Let A be a noetherian ring and N a finitely generated A-module. For an integer r, condition S_r of Serre is the following

S_r : For all $\mathfrak{p} \in \operatorname{Supp}(N)$, $\operatorname{depth}(N_{\mathfrak{p}}) \geq \inf\{r, \dim(N_{\mathfrak{p}})\}$.

It is easy to see that N satisfies condition S_r if and only if, for all $\mathfrak{p} \in \operatorname{Supp}(N)$,

$$\operatorname{depth}(N_{\mathfrak{p}}) < r \quad \Rightarrow \quad N_{\mathfrak{p}} \text{ is Cohen–Macaulay}.$$

Remark that the condition S_1 means that N has *no embedded components*, i.e., $\operatorname{Ass}(N)$ is discrete in $\operatorname{Spec}(A)$.

Example 9.40. If A is a domain and $\operatorname{Supp}(N) = \operatorname{Spec}(A)$, then S_1 means that N is torsion free. Let us also remark that if A is a normal domain, then the A-module A satisfies S_2, 3.9.

Definition 9.41. For $\mathfrak{q} \in \operatorname{Spec}(A)$ define

$$\operatorname{codepth}(N_\mathfrak{q}) = \begin{cases} \dim(N_\mathfrak{q}) - \operatorname{depth}(N_\mathfrak{q}) & \text{if } \mathfrak{q} \in \operatorname{Supp}(N) \\ 0 & \text{otherwise.} \end{cases}$$

Let us recall that for a topological space X, a map $f : X \to \mathbb{Z}$ is called *upper semi-continuous* if for all $x \in X$, there exists a neighborhood U of x such that $f(u) \le f(x)$ for all $u \in U$. Similarly f is called *lower semi-continuous* if for all $x \in X$ there exists a neighborhood U of x such that $f(u) \ge f(x)$ for all $u \in U$.

Proposition 9.42. *Suppose A has a dualizing complex. Then*

$$\mathfrak{q} \mapsto \operatorname{codepth}(N_\mathfrak{q}) : \operatorname{Spec}(A) \to \mathbb{Z}$$

is upper semi-continuous.

Proof. Let $D^{\bullet}$ be a dualizing complex. For $\mathfrak{q} \in \operatorname{Spec}(A)$ we have by 9.13

$$\operatorname{codepth}(N_\mathfrak{q}) = \operatorname{amp}(\operatorname{Hom}_A^{\bullet}(N, D^{\bullet})_\mathfrak{q}).$$

We leave it to the reader to establish that for any bounded complex $X^{\bullet}$ with finitely generated cohomology modules

$$\mathfrak{q} \mapsto \operatorname{amp}(X_\mathfrak{q}^{\bullet}) : \operatorname{Spec}(A) \to \mathbb{Z}$$

is upper semi-continuous. $\square$

Corollary 9.43. *If A has a dualizing complex, then*

$$\mathfrak{q} \mapsto \operatorname{codepth}(N_\mathfrak{q})$$

is bounded.

Proof. For $i \in \mathbb{N}$ let $G_i(N)$ denote the set of $\mathfrak{q}$'s in $\operatorname{Spec}(A)$ for which $\operatorname{codepth}(N_\mathfrak{q}) \le i$. We have

$$\operatorname{Spec}(A) = \bigcup_i G_i(N).$$

The space $\operatorname{Spec}(A)$ is quasi-compact, so finitely many G_i's will cover $\operatorname{Spec}(A)$. $\square$

Remark 9.44. The codepth function can be used to express condition

S_r : for all $n \in \mathbb{N}$ and $\mathfrak{q} \in \mathrm{Supp}(N)$,

$$\mathrm{codepth}(N_\mathfrak{q}) < n \quad \Rightarrow \quad \dim(N_\mathfrak{q}) > n + r.$$

The proof is left to the reader.

Theorem 9.45. *Let N be a finitely generated module whose codepth function is upper semi-continuous. Then for $r \in \mathbb{N}$*

$$\{\mathfrak{p} \in \mathrm{Supp}(N) \mid A_\mathfrak{p}\text{-module } N_\mathfrak{p} \text{ satisfies } S_r\}$$

is an open subset of $\mathrm{Supp}(N)$.

Proof. Put $Z = \mathrm{Supp}(N)$ and for $n \in \mathbb{N}$, $Z_n = \{\mathfrak{q} \in Z \mid \mathrm{codepth}(N_\mathfrak{q}) > n\}$. By assumption this is a closed set. For $\mathfrak{p} \in Z_n$ define

$$\mathrm{codim}_\mathfrak{p}(Z_n, N) = \inf\{\dim(N_\mathfrak{q}) \mid \mathfrak{q} \in Z_n, \mathfrak{q} \subseteq \mathfrak{p}\}.$$

We leave it to the reader to use 9.44 to see that the $A_\mathfrak{p}$-module $N_\mathfrak{p}$ satisfies S_r if and only if for all $n \in \mathbb{N}$

$$\mathrm{codim}_\mathfrak{p}(Z_n, N) > n + r.$$

According to the proof of 9.43, only finitely many Z_n's are nonempty. By 9.46 below

$$\mathfrak{p} \mapsto \mathrm{codim}_\mathfrak{p}(Z_n, N) : Z_n \to \mathbb{Z}$$

is lower semi-continuous. $\qquad\square$

Lemma 9.46. *Let Y be a closed subset of* $\mathrm{Supp}(N)$. *For $\mathfrak{p} \in Y$ define*

$$\mathrm{codim}_\mathfrak{p}(Y, N) = \inf\{\dim(N_\mathfrak{q}) \mid \mathfrak{q} \in Y, \mathfrak{q} \subseteq \mathfrak{p}\}.$$

Then

$$\mathfrak{p} \mapsto \mathrm{codim}_\mathfrak{p}(Y, N) : Y \to \mathbb{Z}$$

is lower semi-continuous.

Proof. Replace N by $A/\mathrm{Ann}(N)$ to se that we may suppose $N = A$. Let $\mathfrak{J}$ be an ideal such that $Y = V(\mathfrak{J})$. Let $\mathcal{Q}$ be the set of minimal prime ideals in $A/\mathfrak{J}$. We have

$$\mathrm{codim}_\mathfrak{p}(Y, A) = \inf\{\dim(A_\mathfrak{q}) \mid \mathfrak{q} \in \mathcal{Q}, \mathfrak{q} \subseteq \mathfrak{p}\}.$$

Now, fix $\mathfrak{p}$ and let $W_\mathfrak{p} = \bigcup_{\mathfrak{q} \in \mathcal{Q}, \mathfrak{q} \not\subseteq \mathfrak{p}} V(\mathfrak{q})$. $W_\mathfrak{p}$ is a closed subset of $\mathrm{Spec}(A)$ not containing $\mathfrak{p}$. The value of the codimension function in $Y - W_\mathfrak{p}$ is bigger or equal $\mathrm{codim}_\mathfrak{p}(Y, A)$. $\qquad\square$

Let us here add a general result on the codepth function.

Proposition 9.47. *Let A denote a noetherian ring and N a finitely generated A-module. For prime ideals $\mathfrak{p}$ and $\mathfrak{q}$ with $\mathfrak{p} \subseteq \mathfrak{q}$*

$$\mathrm{codepth}_{\mathfrak{p}}(N) \leq \mathrm{codepth}_{\mathfrak{q}}(N).$$

Proof. If A has a dualizing complex, this is a consequence of the upper semi-continuity of the codepth function, 9.42. In general we can suppose that A is local with maximal ideal $\mathfrak{q}$ and that $\mathfrak{p} \in \mathrm{Supp}(N)$. Choose a prime ideal $\mathfrak{r}$ of the completion $\hat{A}$ whose retraction to A is $\mathfrak{p}$. This can be done by the first part of 9.48 below applied to $A/\mathfrak{p} \to \hat{A}/\hat{A}\mathfrak{p}$. By the second part of 9.48 applied to the morphism $A_{\mathfrak{p}} \to \hat{A}_{\mathfrak{p}}$, we have

$$\mathrm{codepth}_{\mathfrak{p}}(N) \leq \mathrm{codepth}_{\mathfrak{p}}(\hat{N})$$

and by 9.48 applied to $A \to \hat{A}$

$$\mathrm{codepth}_{\mathfrak{q}}(N) = \mathrm{codepth}_{\mathfrak{q}}(\hat{N}). \qquad \square$$

Lemma 9.48. *Let A and B be noetherian local rings, $f : A \to B$ a flat local homomorphism. Then the inverse image by f of a minimal prime ideal in B is a minimal prime ideal in A. For a finitely generated A-module $M \neq 0$ we have*

$$\dim(M \otimes_A B) = \dim(M) + \dim(k \otimes_A B)$$
$$\mathrm{depth}(M \otimes_A B) = \mathrm{depth}(M) + \mathrm{depth}(k \otimes_A B),$$

where k denotes the residue field of A.

Proof. Let $\mathfrak{q}$ be a prime ideal in B and $\mathfrak{p} = f^{-1}(\mathfrak{q})$. Consider the flat local morphism $A_{\mathfrak{p}} \to B_{\mathfrak{q}}$, then

$$\mathrm{Spec}(B_{\mathfrak{q}}) \to \mathrm{Spec}(A_{\mathfrak{p}})$$

is surjective. In particular if $\mathfrak{q}$ is minimal then $\mathfrak{p}$ is minimal.

The depth formula is a consequence of 8.37.

Let us first prove the dimension formula in case $M = A$. This is done by induction on $\dim(A)$. If $\dim(A) = 0$, $\mathfrak{m}$ is nilpotent and therefore $\mathfrak{m}B$ is nilpotent and whence $\dim(B) = \dim(B/\mathfrak{m}B)$. Suppose $\dim(A) \neq 0$. Choose $a \in \mathfrak{m}$ such that a is outside all the minimal prime ideals of A.

By our first result $f(a)$ is outside all minimal prime ideals of B. Thus $\dim(A/(a)) = \dim(A) - 1$ and $\dim(B/(f(a))) = \dim(B) - 1$. We can now apply the induction hypothesis to the flat morphism $A/(a) \to B/(f(a))$.

In the general case, remark that $\operatorname{Supp}(M \otimes_A B)$ is the inverse image of $\operatorname{Supp}(M)$ by $\operatorname{Spec} A \to \operatorname{Spec} B$. Thus if $\mathfrak{I}$ is an ideal in A such that $V(\mathfrak{I}) = \operatorname{Supp}(M)$, we have $V(f(\mathfrak{I})B) = \operatorname{Supp}(M \otimes_A B)$. Thus it suffices to treat the case $M = A/\mathfrak{I}$. This however is reduced to the previous case by considering the flat morphism $A/\mathfrak{I} \to B/f(\mathfrak{I})B$. $\qquad\square$

9.8   Factorial rings and condition S_r

In this section A denotes a noetherian local ring.

Proposition 9.49. *Suppose A is a factorial local domain with fraction field K. If $M \neq 0$ is a finitely generated module such that M satisfies S_2 and $M \otimes_A K \simeq K$, then M is isomorphic to A.*

Proof. Follows from the proof of 8.28 and 8.29. $\qquad\square$

Theorem 9.50. *Suppose A is a factorial local domain and $D^{\bullet}$ is a normalized dualizing complex. Then $H^0(D^{\bullet}) \simeq A$.*
   Moreover A satisfies condition S_{2+r}, $r > 0$, if and only if

$$H^1(D^{\bullet}) = H^2(D^{\bullet}) = \cdots = H^r(D^{\bullet}) = 0.$$

Proof. Follows from 9.49, 9.52 and 9.54. $\qquad\square$

Corollary 9.51. *With the assumptions of 9.50 suppose A satisfies S_3 and that for all $\mathfrak{p} \in \operatorname{Spec}(A)$*

$$\dim(A_{\mathfrak{p}}) \geq 4 \quad \Rightarrow \quad \operatorname{depth}(A_{\mathfrak{p}}) \geq \tfrac{1}{2}\dim(A_{\mathfrak{p}}) + 1.$$

Then A is Gorenstein.

Proof. We shall prove by induction on r that A satisfies S_r for all $r \in \mathbb{N}$. So suppose A satisfies S_{r-1}. It will now suffice to prove that

$$\operatorname{depth}(A_{\mathfrak{p}}) = r - 1 \quad \Rightarrow \quad A_{\mathfrak{p}} \text{ is Cohen–Macaulay.}$$

We have $\dim(A)p \leq 2r - 4$. Therefore

$$H^i(D_{\mathfrak{p}}^{\bullet}) = 0 \qquad \text{for } i > (2r - 4) - (r - 1) = r - 3.$$

On the other hand, since $A_{\mathfrak{p}}$ satisfies S_{r-1} we have by the theorem

$$H^i(D_{\mathfrak{p}}^{\cdot}) = 0 \qquad \text{for } 1 \le i \le r - 3. \qquad \square$$

In particular a 4-dimensional factorial domain which is S_3 (and has a dualizing complex) is Gorenstein.

In the following A denotes a noetherian local ring of dimension d, and $D^{\cdot}$ is a normalized dualizing complex.

Proposition 9.52. *If $N \ne 0$ is a finitely generated module of dimension n, with the property that, $r \in \mathbb{N}$*

$$\text{Ext}_A^{d-n+1}(N, D^{\cdot}) = \cdots = \text{Ext}_A^{d-n+r}(N, D^{\cdot}) = 0.$$

Then $\text{Ext}_A^{d-n}(N, D^{\cdot})$ satisfies condition S_{r+2}.

Proof. We are going to prove that for all $\mathfrak{p} \in \text{Supp}(\text{Ext}_A^{d-n}(N, D^{\cdot}))$

$$\text{depth}(\text{Ext}_A^{d-n}(N, D^{\cdot})_{\mathfrak{p}}) < r + 2 \;\Rightarrow\; \text{Ext}_A^{d-n}(N, D^{\cdot})_{\mathfrak{p}} \text{ is Cohen–Macaulay.}$$

This has already been done for $r = -1$ in 9.12(3). By induction on r it will suffice to prove that for $\mathfrak{p} \in \text{Supp}(\text{Ext}_A^{d-n}(N, D^{\cdot}))$,

$$\text{depth}(\text{Ext}_A^{d-n}(N, D^{\cdot})_{\mathfrak{p}}) = r + 1 \;\Rightarrow\; \text{Ext}_A^{d-n}(N, D^{\cdot})_{\mathfrak{p}} \text{ is Cohen–Macaulay.}$$

So suppose that $\text{Ext}_A^{r+1}(k_{\mathfrak{p}}, \text{Ext}_A^{d-n}(N, D^{\cdot})_{\mathfrak{p}}) \ne 0$ and moreover that $\text{Ext}_A^i(k_{\mathfrak{p}}, \text{Ext}_A^{d-n}(N, D^{\cdot})_{\mathfrak{p}}) = 0$ for $i \le r$. By the following lemma we have

$$i(\text{Ext}_A^{\cdot}(k_{\mathfrak{p}}, \text{Hom}_A^{\cdot}(N, D^{\cdot})_{\mathfrak{p}})) = d - n + r + 1.$$

On the other hand, by 9.14

$$i(\text{Ext}_A^{\cdot}(k_{\mathfrak{p}}, \text{Hom}_A^{\cdot}(N, D^{\cdot})_{\mathfrak{p}})) = i(\text{Ext}_A^{\cdot}(k_{\mathfrak{p}}, D_{\mathfrak{p}}^{\cdot})) = d - \dim(A/\mathfrak{p}).$$

Thus, $d - n + r + 1 = d - \dim(A/\mathfrak{p})$ or

$$\dim(N) - \dim(A/\mathfrak{p}) = \text{depth}(\text{Ext}_A^{d-n}(N, D^{\cdot})_{\mathfrak{p}}).$$

Since

$$\dim(\text{Ext}_A^{d-n}(N, D^{\cdot})_{\mathfrak{p}}) + \dim(A/\mathfrak{p}) \le \dim(N)$$

we conclude that $\text{Ext}_A^{d-n}(N, D^{\cdot})_{\mathfrak{p}}$ is Cohen–Macaulay. $\qquad \square$

Lemma 9.53. *Let $X^{\cdot}$ be a bounded below complex. Given integers $c \in \mathbb{Z}$ and $r \in \mathbb{N}$ such that*

$$H^i(X^{\cdot}) = 0 \qquad \text{for } i < c \text{ or } c < i \leq c+r.$$

Then for any module K, we have isomorphisms

$$\mathrm{Ext}_A^j(K, H^c(X^{\cdot})) \simeq \mathrm{Ext}_A^{c+j}(K, X^{\cdot}) \qquad \text{for } j \leq r$$

and an exact sequence

$$0 \longrightarrow \mathrm{Ext}_A^{r+1}(K, H^c(X^{\cdot})) \longrightarrow \mathrm{Ext}_A^{c+r+1}(K, X^{\cdot}).$$

Proof. $X^{\cdot}$ is quasi-isomorphic to a complex with $X^i = 0$ for $i < c$. Thus we may assume this. We have a short exact sequence of complexes:

$$
\begin{array}{ccccccccccc}
0 & \longrightarrow & H^c(X^{\cdot}) & \longrightarrow & 0 & \longrightarrow & \cdots & \longrightarrow & 0 & \longrightarrow & 0 & \longrightarrow \\
& & \downarrow & & \downarrow & & & & \downarrow & & \downarrow \\
0 & \longrightarrow & X^c & \longrightarrow & X^{c+1} & \longrightarrow & \cdot s & \longrightarrow & X^{c+r} & \longrightarrow & X^{c+r+1} & \longrightarrow \\
& & \downarrow & & \downarrow & & & & \downarrow & & \downarrow \\
0 & \longrightarrow & Y^c & \longrightarrow & Y^{c+1} & \longrightarrow & \cdots & \longrightarrow & Y^{c+r} & \longrightarrow & Y^{c+r+1} & \longrightarrow
\end{array}
$$

From the exact cohomology sequence follows $H^i(Y^{\cdot}) = 0$ for $i \leq c+r$. Therefore $\mathrm{Ext}_A^i(K, Y^{\cdot}) = 0$ for $i \leq c+r$. The long exact sequence in $\mathrm{Ext}_A^{\cdot}(K, -)$ will give the result. $\qquad\square$

Proposition 9.54. *Let $N \neq 0$ be a finitely generated A-module, which is equidimensional, i.e., $n = \dim(A/\mathfrak{p})$ is the same for all minimal prime ideals $\mathfrak{p} \in \mathrm{Supp}(N)$. Let $r \in \mathbb{N}$ be such that the following condition holds, $d = \dim(A)$, for all $\mathfrak{p} \in \mathrm{Supp}(\mathrm{Ext}_A^{d-n}(N, D^{\cdot}))$*

$$\mathrm{depth}(\mathrm{Ext}_A^{d-n}(N, D^{\cdot})_{\mathfrak{p}}) < r + 2 \quad \Rightarrow \quad N_{\mathfrak{p}} \text{ is Cohen–Macaulay.}$$

Then

$$\mathrm{Ext}_A^{d-n+1}(N, D^{\cdot}) = \cdots = \mathrm{Ext}_A^{d-n+r}(N, D^{\cdot}) = 0.$$

Proof. Let us first remark that $\mathrm{Supp}(\mathrm{Ext}_A^{d-n}(N, D^{\cdot})) = \mathrm{Supp}(N)$ by 9.12(3). Note also if $\mathfrak{q} \in \mathrm{Supp}(N)$, then $N_{\mathfrak{q}}$ is still equidimensional, since A is catenary. Moreover, $d - n = \dim(A_{\mathfrak{q}}) - \dim(N_{\mathfrak{q}})$. Thus we can proceed by induction on $d = \dim(A)$. By the induction hypothesis we may assume that

$$\mathrm{Ext}_A^{d-n+1}(N, D^{\cdot}), \ldots, \mathrm{Ext}_A^{d-n+r}(N, D^{\cdot})$$

all have finite length. We shall assume that the complex $D^{\cdot}$ is minimal, this has by the proof of 9.12 the effect that

$$\operatorname{Hom}_A(N, D^j) = 0, \qquad j < d - n.$$

We are going to apply the acyclicity 9.33 to the complex

$$0 \longrightarrow \operatorname{Ext}_A^{d-n}(N, D^{\cdot}) \longrightarrow \operatorname{Hom}_A(N, D^{d-n})$$
$$\longrightarrow \cdots \longrightarrow \operatorname{Hom}_A(N, D^{d-n+r+1}) \longrightarrow 0.$$

If $\operatorname{depth}(\operatorname{Ext}_A^{d-n}(N, D^{\cdot})) < r + 2$ then N is Cohen–Macaulay and there is nothing to prove. In case $\operatorname{depth}(\operatorname{Ext}_A^{d-n}(N, D^{\cdot})_{\mathfrak{p}}) \geq r + 2$ we have $n = \dim(\operatorname{Ext}_A^{d-n}(N, D^{\cdot})) \geq r + 2$, and whence $d - n + r + 1 < d$. We can now conclude the proof by showing that $\operatorname{Ext}_A^{\cdot}(k, \operatorname{Hom}_A(N, D^j)) = 0$ for $j < d$. Note first $\operatorname{Ext}_A^{\cdot}(k, D^j) = 0$. Choose a free resolution $L_{\cdot} \to N$. This makes $\operatorname{Hom}_A^{\cdot}(L_{\cdot}, D^j)$ an injective resolution of $\operatorname{Hom}_A(N, D^j)$. Therefore

$$\operatorname{Ext}_A^{\cdot}(k, \operatorname{Hom}_A(N, D^j)) = H^{\cdot}(\operatorname{Hom}_A^{\cdot}(L^{\cdot} \otimes_A k, \operatorname{Hom}_A(k, D^j))) = 0. \qquad \square$$

9.9 Condition S'_r

Let A denote a noetherian local ring with maximal ideal $\mathfrak{m}$ and residue field k.

Definition 9.55. Let $r \in \mathbb{Z}$. A finitely generated module N satisfies the condition S'_r if

S'_r : For all $\mathfrak{p} \in \operatorname{Supp}(N)$, $\operatorname{depth}(N_{\mathfrak{p}}) \geq \inf\{r, \dim(N) - \dim(A/\mathfrak{p})\}$.

Remark 9.56. For any $\mathfrak{p} \in \operatorname{Supp}(N)$

$$\dim(N) - \dim(A/\mathfrak{p}) \geq \dim(N_{\mathfrak{p}})$$

thus S'_r implies S_r.

If $\operatorname{Supp}(N)$ is equidimensional and catenary, i.e., for all $\mathfrak{p} \in \operatorname{Supp}(N)$

$$\dim(N) - \dim(A/\mathfrak{p}) = \dim(N_{\mathfrak{p}})$$

then the two conditions are equivalent.

Remark 9.57. To illustrate the differences let us remark that for $N \neq 0$ the two conditions read

S_1: N has no embedded components.

S_1': N has no embedded components and $\mathrm{Supp}(N)$ is equidimensional.

Remark 9.58. A convenient way of rephrasing condition S_r' is the following

S_r' : For all $\mathfrak{p} \in \mathrm{Supp}(N)$, $\mathrm{depth}(N_\mathfrak{p}) \leq r \Rightarrow N_\mathfrak{p}$, is Cohen–Macaulay and $\dim(N_\mathfrak{p}) + \dim(A/\mathfrak{p}) = \dim N$.

In the following we put $d = \dim(A)$ and let $D^{\textstyle\cdot}$ denote a dualizing complex.

Theorem 9.59. *Let $N \neq 0$ be a finitely generated module with $n = \dim(N)$. Given an integer $r \in \mathbb{N}$, then N satisfies condition S_r' if and only if*

$$\dim(\mathrm{Ext}_A^i(N, D^{\textstyle\cdot})) \leq d - i - r$$

for all $d - n < i \leq d$.

Proof. Given $\mathfrak{p} \in \mathrm{Supp}(N)$, let us reformulate the results of section 9.2 for the ring $A_\mathfrak{p}$, the module $N_\mathfrak{p}$ and the (in general not normalized) dualizing complex $D_\mathfrak{p}^{\textstyle\cdot}$

$$\mathrm{Ext}_A^i(N, D^{\textstyle\cdot})_\mathfrak{p} \begin{cases} = 0 & \text{for } i < d - \dim(A/\mathfrak{p}) - \dim(N_\mathfrak{p}) \\ \neq 0 & \text{for } i = d - \dim(A/\mathfrak{p}) - \dim(N_\mathfrak{p}) \\ \neq 0 & \text{for } i = d - \dim(A/\mathfrak{p}) - \mathrm{depth}(N_\mathfrak{p}) \\ = 0 & \text{for } i > d - \dim(A/\mathfrak{p}) - \mathrm{depth}(N_\mathfrak{p}). \end{cases}$$

They all follow from section 9.2 by remarking that $D_\mathfrak{p}^{\textstyle\cdot}[-\dim(A/\mathfrak{p}) + d]$ is a normalized dualizing complex for $A_\mathfrak{p}$.

Let us say that N satisfies condition T_r if

T_r : For all $d - n < i \leq d$, $\dim(\mathrm{Ext}_A^i(N, D^{\textstyle\cdot})) \leq d - i - r$.

Let us first prove that S_r' implies T_r. This is done by induction on r. The first part of 9.12 says that T_0 is always true.

Let $r > 0$ and assume T_{r-1} and S_r'. Given $d - n < i \leq d$ we have to prove

$$\dim(\mathrm{Ext}_A^i(N, D^{\textstyle\cdot})) \leq d - i - r.$$

We are going to prove this by contradiction. So suppose that this inequality does not hold. Then we can find $\mathfrak{p} \in \mathrm{Supp}(\mathrm{Ext}_A^i(N, D^{\boldsymbol{\cdot}}))$ such that $\dim(A/\mathfrak{p}) > d - i - r$. Since we have assumed T_{r-1} we have $\dim(\mathrm{Ext}_A^i(N, D^{\boldsymbol{\cdot}})) \leq d - i - r + 1$. Consequently we have

$$\dim(A/\mathfrak{p}) = d - 1 - r + 1.$$

Since $\mathfrak{p} \in \mathrm{Supp}(\mathrm{Ext}_A^i(N, D^{\boldsymbol{\cdot}})$, we have

$$i \leq d - \dim(A/\mathfrak{p}) - \mathrm{depth}(N_\mathfrak{p}).$$

Now eliminate $\dim(A/\mathfrak{p})$ from these two results to get

$$\mathrm{depth}(N_\mathfrak{p}) \leq r - 1.$$

By condition S'_r this implies

$$N_\mathfrak{p} \text{ is Cohen–Macaulay and } \mathrm{depth}(N_\mathfrak{p}) = n - \dim(A/\mathfrak{p}).$$

In general for $j = d - \dim(N_\mathfrak{p}) - \dim(A/\mathfrak{p})$, $\mathrm{Ext}_A^j(N, D^{\boldsymbol{\cdot}})_\mathfrak{p} \neq 0$. Whence we have $\mathrm{Ext}_A^{d-n}(N, D^{\boldsymbol{\cdot}})_\mathfrak{p} \neq 0$. Since $N_\mathfrak{p}$ is Cohen–Macaulay, this gives

$$\mathrm{Ext}_A^i(N, D^{\boldsymbol{\cdot}})_\mathfrak{p} = 0 \qquad \text{for } i > d - n$$

a contradiction.

Let us now prove that T_r implies S'_r. We proceed by induction. Condition S'_0 is vacuous.

Let $r > 0$ and assume T_r and S'_{r-1}. Let $\mathfrak{p} \in \mathrm{Supp}(N)$, the problem is to prove

$$N_\mathfrak{p} \text{ is Cohen–Macaulay and } \mathrm{depth}(N_\mathfrak{p}) = \dim(N) - \dim(A/\mathfrak{p})$$

under the assumption $\mathrm{depth}(N_\mathfrak{p}) = r - 1$. This implies

$$\mathfrak{p} \in \mathrm{Supp}(\mathrm{Ext}_A^i(N, D^{\boldsymbol{\cdot}})), \qquad i = d - \dim(A/\mathfrak{p}) - r + 1.$$

Suppose first $i \leq d - n$. That means $r - 1 \geq n - \dim(A/\mathfrak{p})$ or

$$\mathrm{depth}(N_\mathfrak{p}) \geq \dim(N) - \dim(A/\mathfrak{p}).$$

However $\dim(N) - \dim(A/\mathfrak{p}) \geq \dim(N_\mathfrak{p})$, so the proof is done in this case.

Suppose second $i > d - n$. By assumption $\dim(\mathrm{Ext}_A^i(N, D^{\boldsymbol{\cdot}})) \leq d - i - r$, in particular

$$\dim(A/\mathfrak{p}) \leq d - i - r = \dim(A/\mathfrak{p}) - 1$$

so the second case does not occur. $\qquad\square$

Corollary 9.60. *Let $r \in \mathbb{N}$. Then a finitely generated A-module N satisfies condition S'_r if and only if the $\hat{A}$-module $\hat{N}$ satisfies condition S'_r.*

Proof. Follows from 9.59 and 8.38.   □

Proposition 9.61. *Let $\mathfrak{p}$ be a prime ideal in A. Then $\hat{A}/\hat{A}\mathfrak{p}$ is equidimensional without embedded components.*

Proof. $\hat{A}/\hat{A}\mathfrak{p}$ is the completion of $A/\mathfrak{p}$. This local ring satisfies condition S'_1 and has a dualizing complex.   □

9.10   Specialization of Poincaré series

Let A be a noetherian ring and $\mathfrak{p}$ a prime ideal. For a bounded complex $X^{\cdot}$ with finitely generated cohomology modules, we define Poincaré series, compare 8.7,

$$\mu_{\mathfrak{p}}(X^{\cdot}, t) = \sum_i \mu_{\mathfrak{p}}^i(X^{\cdot}) t^i = \sum_i \operatorname{rank}_{k_{\mathfrak{p}}}(\operatorname{Ext}_{A_{\mathfrak{p}}}^i(k_{\mathfrak{p}}, X_{\mathfrak{p}}^{\cdot})) t^i = \mu(X_{\mathfrak{p}}^{\cdot}, t)$$

$$\beta^{\mathfrak{p}}(X^{\cdot}, t) = \sum_i \beta_i^{\mathfrak{p}}(X^{\cdot}) t^i = \sum_i \operatorname{rank}_{k_{\mathfrak{p}}}(\operatorname{Tor}_i^{A_{\mathfrak{p}}}(k_{\mathfrak{p}}, X_{\mathfrak{p}}^{\cdot})) t^i = \beta(X_{\mathfrak{p}}^{\cdot}, t).$$

For $\mathfrak{p} \subseteq \mathfrak{q}$ two prime ideals, we are going to compare the Poincaré series

$$\beta^{\mathfrak{p}}(X^{\cdot}, t) \qquad \text{with } \beta^{\mathfrak{q}}(X^{\cdot}, t)$$
$$\mu_{\mathfrak{p}}(X^{\cdot}, t) \qquad \text{with } \mu_{\mathfrak{q}}(X^{\cdot}, t).$$

Proposition 9.62. *With the notation above let $r = \dim((A/\mathfrak{p})_{\mathfrak{q}})$. Then for all $i \in \mathbb{Z}$*

$$\beta_i^{\mathfrak{p}}(X^{\cdot}) \leq \beta_i^{\mathfrak{q}}(X^{\cdot})$$
$$\mu_{\mathfrak{p}}^i(X^{\cdot}) \leq \mu_{\mathfrak{q}}^{i+r}(X^{\cdot}).$$

Proof. We may assume A is local with maximal ideal $\mathfrak{q}$ and residue field k. By 7.12 below we can choose $L^{\cdot} \to X^{\cdot}$ a quasi-isomorphism, where $L^{\cdot}$ is a complex of finitely generated free modules such that $L^{\cdot} \otimes_A k$ has a zero differential. We have

$$\operatorname{rank}_A(L_i) = \beta_i^{\mathfrak{q}}.$$

It follows from 9.63 below, applied to $A_{\mathfrak{p}}$ and $L_{\cdot \mathfrak{p}}$ that

$$\beta_i^{\mathfrak{p}} = \operatorname{rank}_{k_{\mathfrak{p}}} H_i(L. \otimes_A k_{\mathfrak{p}}) \leq \operatorname{rank}_A(L_i) = \beta_i^{\mathfrak{q}}.$$

First assume that A has a dualizing complex $D^{\cdot}$, we shall assume

$$\mu_{\mathfrak{q}}(D^{\cdot}, t) = 1.$$

This implies by 8.3

$$\mu_{\mathfrak{q}}(X^{\cdot}, t) = \beta^{\mathfrak{q}}(\operatorname{Hom}_A^{\cdot}(X^{\cdot}, D^{\cdot}), t).$$

Note we have

$$\mu_{\mathfrak{p}}(D^{\cdot}, t) = t^{-r}.$$

Whence by 8.3

$$\mu_{\mathfrak{p}}(D^{\cdot}, t) = t^r \beta^{\mathfrak{q}}(\operatorname{Hom}_A^{\cdot}(X^{\cdot}, D^{\cdot}), t).$$

Thus the result follows by applying our first result to $\operatorname{Hom}_A^{\cdot}(X^{\cdot}, D^{\cdot})$. Let $\hat{\mathfrak{q}}$ denote the maximal ideal of $\hat{A}$. We have $\hat{A}/\hat{A}\mathfrak{p} = (A/\mathfrak{p})\hat{\ }$, whence

$$\dim(\hat{A}/\hat{A}\mathfrak{p}) = \dim(A/\mathfrak{p}) = r$$

by 4.15. Choose a prime ideal $\bar{\mathfrak{p}}$ in $\hat{A}/\hat{A}\mathfrak{p}$ such that

$$\dim(\hat{A}/\bar{\mathfrak{p}}) = r.$$

It is clear that $\bar{\mathfrak{p}}$ is minimal therefore by 9.44 the contraction to A is $\mathfrak{p}$.
We can now apply 8.37 to the local morphism $A_{\mathfrak{p}} \to \hat{A}_{\bar{\mathfrak{p}}}$

$$\mu_{\bar{\mathfrak{p}}}(X^{\cdot} \otimes_A \hat{A}, t) = \mu_{\mathfrak{p}}(X^{\cdot}, t)\mu_{\bar{\mathfrak{p}}}(\hat{A}_{\bar{\mathfrak{p}}}/\mathfrak{p}\hat{A}_{\bar{\mathfrak{p}}}, t).$$

The last series is calculated relative to the artinian ring $\hat{A}_{\bar{\mathfrak{p}}}/\mathfrak{p}\hat{A}_{\bar{\mathfrak{p}}}$. In particular this series has nonzero constant term. It follows that

$$\mu_{\bar{\mathfrak{p}}}^i(X^{\cdot} \otimes_A \hat{A}) \geq \mu_{\mathfrak{p}}^i(X^{\cdot})$$

clearly

$$\mu_{\hat{\mathfrak{q}}}^i(X^{\cdot} \otimes_A \hat{A}) = \mu_{\mathfrak{q}}^i(X^{\cdot}). \qquad \square$$

In the following A denotes a noetherian local ring with residue field k.

Lemma 9.63. *Let $X.$ be a bounded above complex of finitely generated free modules. For all $i \in \mathbb{Z}$*

$$\operatorname{rank}_k(H_i(X. \otimes_A k)) \leq \operatorname{rank}_A(X_i).$$

Proof. By 7.12 choose a quasi-isomorphism $f : L. \to X.$, where $L^{\cdot}$ is a complex of finitely generated free modules such that $L^{\cdot} \otimes_A k$ has a zero differential. Again by 7.12, f has a homotopy inverse $g : X. \to L.$. Now consider the situation after tensoring with k. The composite

$$L. \otimes_A k \xrightarrow{\ f \otimes 1\ } X. \otimes_A k \xrightarrow{\ g \otimes 1\ } L. \otimes_A k$$

is homotopic to the identity. But since $L. \otimes_A k$ has zero differentials it follows that $g \circ f \otimes 1$ is the identity. In particular it follows from Nakayama's lemma that $g_i : X_i \to L_i$ is an epimorphism, therefore rank $L_i \leq$ rank X_i, on the other hand,

$$\mathrm{rank}_k(H_i(X. \otimes_A k) = \mathrm{rank}_k(H_i(L. \otimes_A k)) = \mathrm{rank}_A(L_i). \qquad \square$$

Chapter 10

Intersection Multiplicities

10.1 Introduction to Serre's conjectures

Let A denote a regular local ring. The main purpose of this chapter is to prove the following theorem.

Theorem 10.1. *Let $M \neq 0$ and $N \neq 0$ be finitely generated A-modules such that $M \otimes_A N$ has finite length. Then*

$$\dim(M) + \dim(N) \leq \dim(A).$$

For M and N finitely generated modules such that $M \otimes_A N$ has finite length, the modules $\mathrm{Tor}_i^A(M, N)$, $i \in \mathbb{N}$ all have finite length, since

$$\mathrm{Supp}(\mathrm{Tor}_i^A(M, N)) \subseteq \mathrm{Supp}(M) \cap \mathrm{Supp}(N) = \mathrm{Supp}(M \otimes_A N).$$

Definition 10.2. For M and N finitely generated modules such that $M \otimes_A N$ has finite length, define the *intersection multiplicity*

$$\chi_A(M, N) = \sum_i (-1)^i \ell_A(\mathrm{Tor}_i^A(M, N)).$$

(In case A is a general noetherian local ring, this makes sense when M or N has a finite free resolution.)

J.–P. Serre has made the following two conjectures.

Let $M \neq 0$ and $N \neq 0$ be finitely generated A-modules such that $M \otimes_A N$ has finite length. Then

Conjecture 10.3.

$$\dim(M) + \dim(N) = \dim(A) \quad \Rightarrow \quad \chi_A(M, N) > 0. \qquad \square$$

Conjecture 10.4.

$$\dim(M) + \dim(N) < \dim(A) \quad \Rightarrow \quad \chi_A(M, N) = 0. \qquad \Box$$

1980: The conjectures are still open in general, but we are going to prove (following Serre) the theorem below

Theorem 10.5. *The conjectures 10.3 and 10.4 hold in the case $A = R[[X_1, \ldots, X_d]]$, where R is a field or a discrete valuation ring.*

Let us at once remark that the proof of 10.1 in general proceeds by first proving 10.1, 10.3 and 10.4 in the case where $A = R[[X_1, \ldots, X_d]]$ as in 10.5, and then use Cohen's structure theorem, 4.17 or section 10.7, to deduce 10.1 in general.

The proof of 10.5 in case where R is a field is done in section 10.5 by a method known as *reduction to the diagonal*, which reduces the problem to the case where $M = A/(a_1, \ldots, a_r)$ where $a. = (a_1, \ldots, a_r)$ is a regular sequence, i.e., a_i is a nonzero divisor for $A/(a_1, \ldots, a_{r-1})$, $i = 1, \ldots, r$. This particular case can be handled in general, sections 10.2 and 10.3, by means of the Koszul complex. The proof of 10.5 in case R is a discrete valuation ring runs, section 10.6, as follows. The case where M and N are R torsion free modules can be handled as if R was a field. In general we can find a filtration of M and of N, whose factors are either R torsion free or annihilated by the maximal ideal of R. The last case can then be handled by appealing to the case where R is a field, by means of a *projection formula*, section 10.4.

Let us conclude this section by a lemma, which shows that Serre's conjectures are additive in the following sense.

Lemma 10.6. *Let M be a nonzero finitely generated module and*

$$0 \longrightarrow P \longrightarrow N \longrightarrow Q \longrightarrow 0$$

a short exact sequence of nonzero finitely generated modules. Suppose $M \otimes_A N$ has finite length and that M, P and M, Q satisfy 10.1 (resp. 10.1, 10.3 and 10.4) then M, N satisfy 10.1 (resp. 10.1, 10.3 and 10.4).

Proof. We have

$$\dim(N) = \sup\{\dim(P), \dim(Q)\},$$

since $\mathrm{Supp}(N) = \mathrm{Supp}(P) \cup \mathrm{Supp}(Q)$. This immediately proves the statement relative to 10.1. Suppose M, P and M, Q satisfy 10.1, 10.3 and 10.4.

It follows that $\chi_A(M, P) \geq 0$ and $\chi_A(M, Q) \geq 0$. We have

$$\chi_A(M, N) = \chi_A(M, P) + \chi_A(M, Q).$$

If $\dim(M) + \dim(N) = \dim(A)$, then either the equality $\dim(N) = \dim(P)$ or $\dim(N) = \dim(Q)$. We have $\chi_A(M, P) > 0$ or $\chi_A(M, Q) > 0$ and whence $\chi_A(M, N) > 0$. If $\dim(M) + \dim(N) < \dim(A)$, we have

$$\dim(M) + \dim(P) = \dim(M) + \dim(Q) = \dim(A)$$

and consequently $\chi_A(M, P) = \chi_A(M, Q) = 0$ and whence $\chi_A(M, N) = 0$. $\square$

Remark 10.7. If the completion $\hat{A}$ satisfies 10.1, 10.3 and 10.4, then A satisfies 10.1, 10.3 and 10.4 as it follows from section 4.4.

10.2 Filtration of the Koszul complex

Let A denote a ring, $a. = (a_1, \ldots, a_r)$ a sequence of elements of A, $\mathfrak{I}$ the ideal generated by $a.$ and $K. = K.(a.)$ the Koszul complex section 2.7.

Lemma 10.8. *For a finitely generated module M, the support of the Koszul complex is*

$$\mathrm{Supp}(K. \otimes_A M) = \mathrm{Supp}(M/\mathfrak{I}M).$$

Moreover, $H.(K. \otimes_A M)$ is annihilated by $\mathfrak{I} + \mathrm{Ann}(M)$.

Proof. We have $H^0(K. \otimes_A M) \simeq M/\mathfrak{I}M$, whence

$$\mathrm{Supp}(M/\mathfrak{I}M) \subseteq \mathrm{Supp}(K. \otimes_A M).$$

On the other hand, if $\mathfrak{p} \notin \mathrm{Supp}(M/\mathfrak{I}M)$ then either $M_{\mathfrak{p}} = 0$ or $\mathfrak{I} \not\subseteq \mathfrak{p}$. In the first case $(K. \otimes_A M)_{\mathfrak{p}} = 0$. In the second, $K._{\mathfrak{p}}$ is homotopic equivalent to zero, since $K.$ is the tensor product of the elementary complexes $A \xrightarrow{a_i} A$ where we can assume $a_1 \notin \mathfrak{p}$ and therefore $A \xrightarrow{a_1} A$ is homotopic equivalent to zero.

Next we see that $H.(K. \otimes_A M)$ is annihilated by $\mathfrak{I}$. Consider the polynomial ring $A[X_1, \ldots, X_r]$ and put $L. = K.(X_1, \ldots, X_r)$. According to 2.39 $L.$ is a free resolution of the $A[X_1, \ldots, X_r]$-module A, where the X_i's act trivial on A. Consider M as an $A[X_1, \ldots, X_r]$-module where X_i acts as multiplication by a_i. It is easily seen that

$$L \otimes_{A[X_1, \ldots, X_r]} M \simeq K. \otimes_A M$$

whence
$$H_i(K. \otimes_A M) = \mathrm{Tor}_i^{A[X_1,\ldots,X_r]}(A, M).$$

The last group is manifestly annihilated by $\mathfrak{J}$.   □

We shall now introduce a *filtration* of $K.$. Let $F^s K.$ denote the subcomplex, $s \in \mathbb{Z}$

$$\cdots \longrightarrow \mathfrak{J}^{s-2}K_2 \longrightarrow \mathfrak{J}^{s-1}K_1 \longrightarrow \mathfrak{J}^s K_0 \longrightarrow 0.$$

We have $F^{s+1}K. \subseteq F^s K.$ and $F^s K. = K.$ for $s \ll 0$. Moreover $\mathfrak{J}F^s K. \subseteq F^{s+1}K.$ and
$$\mathfrak{J}F^s K. = F^{s+1}K., \qquad s \geq 0.$$

For an A-module M we filter $K. \otimes_A M$ by

$$F^s(K. \otimes_A M) = \mathrm{Im}(F^s K. \otimes_A M \to K. \otimes_A M).$$

Proposition 10.9. *If A is noetherian then for any finitely generated A-module M,*
$$K. \otimes_A M \to K. \otimes_A M / F^s(K. \otimes_A M)$$
is a quasi-isomorphism for $s \gg 0$.

Proof. Let $\mathrm{gr}_F(K. \otimes_A M)$ denote the complex of graded modules associated with the complex of filtered modules $K.$. We have

$$\mathrm{gr}_F(K. \otimes_A M) = \oplus_s F^s(K. \otimes_A M)/F^{s+1}(K. \otimes_A M).$$

We are going to prove that

$$H.(F^s(K. \otimes_A M)/F^{s+1}(K. \otimes_A M)) = 0$$

for $s \gg 0$. The complex $\mathrm{gr}_F(K. \otimes_A M)$ is a complex of graded $\mathrm{gr}_{\mathfrak{J}}(A)$-modules. $\mathrm{gr}_{\mathfrak{J}}(A)$ is noetherian, 1.6, and $\mathrm{gr}_F(K. \otimes_A M)$ is a bounded complex of finitely generated $\mathrm{gr}_{\mathfrak{J}}(A)$-modules, 1.7. We have

$$H_i(\mathrm{gr}_F(K. \otimes_A M)) = \oplus_s H_i(F^s(K. \otimes_A M)/F^{s+1}(K. \otimes_A M)).$$

Thus it will suffice to prove that $H_i(\mathrm{gr}_F(K. \otimes_A M))$ is finitely generated as an $A/\mathfrak{J}$-module. Notice

$$\mathrm{gr}_F(K. \otimes_A M) = \mathrm{gr}_F(K.) \otimes_{\mathrm{gr}_{\mathfrak{J}}(A)} \mathrm{gr}_{\mathfrak{J}}(M).$$

We have

$$\mathrm{Supp}(\mathrm{gr}_F(K. \otimes_A M)) = \mathrm{Supp}(\mathrm{gr}_F(K.)) \cap \mathrm{Supp}(\mathrm{gr}_{\mathfrak{J}}(M))$$

and are going to prove that

$$\mathrm{Supp}(\mathrm{gr}_F(K.)) = V(\mathrm{gr}_{\mathfrak{J}}(A)^+).$$

From this it will follow that $H.(\mathrm{gr}_F(K. \otimes_A M))$ is annihilated by a power of $\mathrm{gr}_{\mathfrak{J}}(A)^+$. Let $\bar{a}_i$ denote the residue class in $\mathfrak{J}/\mathfrak{J}^2$. A straightforward direct inspection shows us that $\mathrm{gr}_F(K.)$ is the Koszul complex relative to $\bar{a}_1, \ldots, \bar{a}_r \in \mathrm{gr}_{\mathfrak{J}}(A)$. Conclusion by 10.10 below. $\qquad \square$

By a *filtration* of a complex $X.$ we will understand a decreasing sequence $F = (F^s X.)_{s \in \mathbb{Z}}$ of subcomplexes such that $F^s X. = X.$ for $s \ll 0$. If $\mathfrak{J}$ is an ideal of A we call F an *$\mathfrak{J}$-filtration* if $\mathfrak{J} F^s X. \subseteq F^{s+1} X.$ for all $s \in \mathbb{Z}$. An $\mathfrak{J}$-*stable filtration* of $X.$ is an $\mathfrak{J}$-filtration $F^\cdot$ of $X.$ such that $\mathfrak{J} F^s X. = F^{s+1} X.$ for $s \gg 0$.

Lemma 10.10. *Let A be a noetherian ring, $\mathfrak{J}$ an ideal of A and $(X., \partial.)$ a bounded complex of finitely generated modules. Given a $\mathfrak{J}$-stable filtration F of $X.$ and suppose that*

(1) $\mathrm{Supp}(X.) \subseteq V(\mathfrak{J})$,

(2) $H.(F^s X./F^{s+1} X.) = 0$ *for $s \gg 0$.*

Then

$$X. \to X./F^s X.$$

is a quasi-isomorphism for $s \gg 0$.

Proof. The assumption (2) says that $F^{s+1} X. \to F^s X.$ is a quasi-isomorphism for $s \gg 0$. We are going to prove that there exists $s_0 \in \mathbb{N}$ such that $F^s X. \to F^{s-s_0} X$ induces the zero map on homology for $s \gg 0$. The combination of these two statements will imply that $H.(F^s X.) = 0$ for $s \ll 0$ and thereby that $X. \to X./F^s X.$ is a quasi-isomorphism for $s \gg 0$. The condition (1) is equivalent to the existence of a $s_1 \in \mathbb{N}$ such that $\mathfrak{J}^{s_1} H.(X.) = 0$. By the Artin–Rees lemma there exists $s_2 \in \mathbb{N}$ such that

$$F^s X. \cap \mathrm{Ker}(\partial.) = \mathfrak{J}^{s-s_2}(F^{s_2} X. \cap \mathrm{Ker}(\partial.))$$

thus (let $s_3 \in \mathbb{Z}$ be such that $F^{s_3} X. = X.$)

$$F^s X. \cap \mathrm{Ker}(\partial.) \subseteq \mathfrak{I}^{s-s_2} \mathrm{Ker}(\partial.)$$
$$\subseteq \mathfrak{I}^{s-s_2-s_1} \partial.(X.)$$
$$\subseteq \partial.(F^{s-s_1-s_2-+s_3} X.).$$

In conclusion $F^s X. \to F^{s-s_1-s_2+s_3} X.$ induces zero on homology.   □

In case A is local we can improve 10.10 as follows.

Lemma 10.11. *Let A be a noetherian local ring, $\mathfrak{I}$ an ideal contained in the maximal ideal of A and $X.$ a bounded complex of finitely generated modules equipped with an $\mathfrak{I}$-stable filtration F. Then*

$$X. \to X./F^s X.$$

is a quasi-isomorphism for $s \gg 0$ if and only if

$$H.(F^s X./F^{s+1} X.) = 0$$

for $s \gg 0$. If this is the case, then some power of $\mathfrak{I}$ annihilates $H.(X.)$.

Proof. Assume $X. \to X./F^s X.$ is a quasi-isomorphism for $s \gg 0$. Then

$$X./F^{s+1} X. \to X./F^s X.$$

is a quasi-isomorphism and whence $H.(F^s X./F^{s+1} X.) = 0$ for $s \gg 0$. To prove the converse let us first remark that if we put

$$F^s H.(X.) = \mathrm{Im}(H.(F^s X.) \to H.(X.)),$$

we obtain an $\mathfrak{I}$-filtration of $H.(X.)$. This filtration is $\mathfrak{I}$-stable, the filtration is the image by $\mathrm{Ker}(\partial.) \to H.(X.)$ of the filtration $\mathrm{Ker}(\partial.) \cap F^\cdot X.$. This filtration is $\mathfrak{I}$-stable by the Artin–Rees lemma. By 1.35 any $\mathfrak{I}$-stable filtration is separated, in particular

$$\bigcap_{s \in \mathbb{Z}} F^s H.(X.) = 0.$$

The assumption $H.(F^s X./F^{s+1} X.) = 0$ for $s \gg 0$ implies $F^s X. = F^{s+1} X.$ for $s \gg 0$. Thus $F^s X. = 0$ for $s \gg 0$. From this follows that some power of $\mathfrak{I}$ annihilates $H.(X.)$. We can now conclude by 10.10.   □

10.3 Euler characteristic of the Koszul complex

In this section A denotes a noetherian local ring, $a. = (a_1, \ldots, a_r)$ a sequence of elements in the maximal ideal $\mathfrak{m}$ of A, $\mathfrak{J}$ the ideal generated by $a.$ and $K.(a.)$ the Koszul complex relative to $a..$

Let $M \neq 0$ be a finitely generated A-module such that $M/\mathfrak{J}M$ has finite length. This implies that $M/\mathfrak{J}^s M$ has finite length for all $s \in \mathbb{N}$, since

$$\mathrm{Supp}(M/\mathfrak{J}^s M) = \mathrm{Supp}(M) \cap V(\mathfrak{J}), \qquad s > 0.$$

It follows that the image of $\mathfrak{J}$ in $A/\mathrm{Ann}(M)$ has finite colength and therefore by 1.36

$$s \mapsto \ell_A(M/\mathfrak{J}^s M)$$

is a polynomial with coefficients in $\mathbb{Q}$ for $s \gg 0$. Let $\chi_{\mathfrak{J}}(M, t)$ denote that polynomial. It follows from 4.15 and the proof of 1.36 that

$$\deg \chi_{\mathfrak{J}}(M, t) = \dim(M).$$

Put $c = \dim(M)$ and write

$$\chi_{\mathfrak{J}}(M, t) = e_{\mathfrak{J}}(M) \binom{t}{c} + \text{ lower terms}.$$

It follows from 1.27 that $e_{\mathfrak{J}}(M)$ is a strictly positive integer, the *multiplicity* of $\mathfrak{J}$ in M. Recall that for a complex $X.$ with finite length cohomology we have the *Euler-characteristic*, compare 9.39,

$$\chi(X.) = \sum (-1)^i \ell_A(H_i(X.)).$$

Theorem 10.12. *Let $M \neq 0$ be a finitely generated A-module such that $M \otimes_A A/\mathfrak{J}$ has finite length. Then $\dim(M) \leq r$ and*

$$\chi(K. \otimes_A M) = \begin{cases} e_{\mathfrak{J}}(M) & \text{if } \dim(M) = r \\ 0 & \text{if } \dim(M) < r. \end{cases}$$

Proof. The assertion that $\dim(M) \leq r$ follows from 1.41 applied to the factor ring $A/\mathrm{Ann}(M)$.

Consider the filtration F of $K. \otimes_A M$ introduced in the previous section.

$$K. \otimes_A M \to K. \otimes_A M/F^s(K. \otimes_A M)$$

is a quasi-isomorphism for $s \gg 0$, 10.9. We have

$$K_i \otimes_A M / F^s(K_i \otimes_A M) \simeq \binom{r}{i} M / \mathfrak{J}^{s-i}.$$

In particular $K. \otimes_A M / F^s(K. \otimes_A M)$, has finite length for $s > r$. Whence for $s \gg 0$

$$\chi(K. \otimes_A M) = \Sigma_i (-1)^i \binom{r}{i} \ell_A(M / \mathfrak{J}^{s-i})$$

$$= \Sigma_i (-1)^i \binom{r}{i} \chi_{\mathfrak{J}}(M, s - i)$$

$$= \Delta^r \chi_{\mathfrak{J}}(M, t)|_{t=s}$$

where Δ is the operator on $\mathbb{Q}[t]$ given by

$$\Delta P(t) = P(t + 1) - P(t).$$

The operator Δ lowers the degree by one and $\Delta\binom{t}{r} = \binom{t}{r-1}$, 1.26.   □

Remark 10.13. With the notation of 10.12 suppose $\dim(M) = r$. Then for any sequence $n_1, \ldots, n_r$ of strictly positive integers,

$$e_{(a_1^{n_1}, \ldots, a_r^{n_r})}(M) = n_1 \ldots n_r e_{(a_1, \ldots, a_r)}(M).$$

This follows from an $\mathfrak{J}$-stable filtration of the following Koszul complex $K.(a_1^{n_1}, \ldots, a_r^{n_r}) \otimes_A M$.

Example 10.14. Let ∂ denote a $(k + 1) \times k$-matrix with coefficients in $\mathfrak{m}$. Let ∂_i denote the determinant of the matrix obtained from ∂ by deleting the i'th row. Put $\partial' = (\partial_1, -\partial_2, \ldots)$ and let $L.$ denote the complex

$$0 \longrightarrow A^k \overset{\partial}{\longrightarrow} A^{k+1} \overset{\partial'}{\longrightarrow} A \longrightarrow 0.$$

Let $\mathfrak{J}$ denote the ideal generated by the coefficients of ∂'. We have

$$\mathrm{Supp}(L.) = V(\mathfrak{J}).$$

Let $\mathfrak{I}$ denote the ideal generated by the coefficients of ∂ and let $F = (F^s L.)_{s \in \mathbb{Z}}$ denote the filtration of $L.$ given by

$$F^s L = \quad 0 \longrightarrow \mathfrak{I}^{s-k-1} A^k \longrightarrow \mathfrak{I}^{s-k} A^{k+1} \longrightarrow \mathfrak{I}^s \longrightarrow 0.$$

This is an $\mathfrak{J}$-stable filtration. For a finitely generated module M we are going to prove that

$$L_. \otimes_A M \to L_. \otimes_A M/F^s(L_. \otimes_A M)$$

is a quasi-isomorphism for $s \gg 0$, if and only if $\mathfrak{J}^q \mathfrak{J} M = \mathfrak{J}^{q+k} M$ for some $q \in \mathbb{N}$. To see this we put $G = \operatorname{gr}_{\mathfrak{J}}(A)$ and let d denote the matrix whose coefficients are the residue classes in $\mathfrak{J}/\mathfrak{J}^2$ of the coefficients of ∂. Let d' denote the matrix obtained from d in the same way ∂' was obtained from ∂. We have

$$\operatorname{gr}_F(L_.) = \quad 0 \longrightarrow G^k \xrightarrow{\ d\ } G^{k+1} \xrightarrow{\ d'\ } G \longrightarrow 0$$

and $\operatorname{gr}_F(L_. \otimes_A M) = \operatorname{gr}_F(L_.) \otimes_G \operatorname{gr}_{\mathfrak{J}}(M)$. Let $\bar{\mathfrak{J}}$ denote the ideal of maximal minors of d. We have

$$\operatorname{Supp}(\operatorname{gr}_F(L_.)) = V(\bar{\mathfrak{J}})$$

and whence

$$\operatorname{Supp}(\operatorname{gr}_F(L_. \otimes_A M)) = \operatorname{Supp}(\operatorname{gr}_{\mathfrak{J}}(M)/\bar{\mathfrak{J}} \operatorname{gr}_{\mathfrak{J}}(M)).$$

Thus by 10.4 $L_. \otimes_A M \to L_. \otimes_A M/F^s(L_. \otimes_A M)$ is a quasi-isomorphism for $s \gg 0$ if and only if

$$\mathfrak{J}\mathfrak{J}^s M + \mathfrak{J}^{s+k+1} M = \mathfrak{J}^{s+k} M$$

for $s \gg 0$, or what is the same, the identity above holds for one $s_0 \in \mathbb{N}$. Lemma 10.15 below follows $\mathfrak{J}^{s_0} \mathfrak{J} M = \mathfrak{J}^{s_0+k} M$.

In the rest of this example, we assume that there exists a $q \in \mathbb{N}$ such that $\mathfrak{J}^q \mathfrak{J} M = \mathfrak{J}^{q+k} M$. This is for example the case if ∂ has the form

$$\begin{pmatrix} x & & & \\ y & x & & \\ & y & x & \\ & & \ddots & \\ & & & y \end{pmatrix}$$

Now suppose that $M \neq 0$ is such that for $M/\mathfrak{J}M$ has finite length. Then for $s \gg 0$ we have

$$\chi(L_. \otimes_A M) = \chi_{\mathfrak{J}}(M, s) - (k+1)\chi_{\mathfrak{J}}(M, s-k) + k\chi_{\mathfrak{J}}(M, s-k-1).$$

The operator $\Gamma_k : \mathbb{Q}[t] \to \mathbb{Q}[t]$ given by

$$\Gamma_k(P(t)) = P(t) - (k+1)P(t-k) + kP(t-k-1)$$

lowers the degree by two and $\Gamma_k(\binom{t}{2}) = \binom{k+1}{2}$.

Since $\chi(L. \otimes_A M) = \Gamma_k(\chi_{\mathfrak{I}}(M,t))|_{t=s}$, for $s \gg 0$, it follows that $\deg \chi_{\mathfrak{I}}(M,t) \leq 2$, and therefore $\dim(M) \leq 2$ and

$$\chi(L. \otimes_A M) = \begin{cases} 0 & \text{if } \dim(M) < 2 \\ \binom{k+1}{2}e_{\mathfrak{I}}(M) & \text{if } \dim(M) = 2. \end{cases}$$

Let us finally prove that if $\dim(M) = 2$ then

$$e_{\mathfrak{I}}(M) = k^2 e_{\mathfrak{I}}(M).$$

To prove this let us first notice that $e_{\mathfrak{I}}(M) = e_{\mathfrak{I}}(\mathfrak{I}^q M)$ and $e_{\mathfrak{I}}(M) = e_{\mathfrak{I}}(\mathfrak{I}^q M)$ as it follows from 1.39. It follows easily from $\mathfrak{I}\mathfrak{I}^q M = \mathfrak{I}^{q+k} M$ that $\chi_{\mathfrak{I}}(\mathfrak{I}^q M, t) = \chi_{\mathfrak{I}}(\mathfrak{I}^q M, kt)$. By assumption $\dim(M/\mathfrak{I}M) = 0$, so $\dim(\mathfrak{I}^q M) = \dim(M) = 2$ and we get by the calculation above $e_{\mathfrak{I}}(\mathfrak{I}^q M) = k^2 e_{\mathfrak{I}}(\mathfrak{I}^q M)$.

Lemma 10.15. *Let $\mathfrak{I}$ be an ideal contained in the maximal ideal of A. Given a finitely generated A-module M and two $\mathfrak{I}$-stable filtrations F_1 and F_2 of M, such that for all $s \in \mathbb{Z}$, $F_1^s M \subseteq F_2^s M$. If there exists s_0 such that*

$$F_1^s M + F_2^{s+1} M = F_2^s M \qquad \text{for } s \geq s_0,$$

then $F_1^s M = F_2^s M$, for $s \geq s_0$.

Proof. Put $N = M/F_1^{s_0} M$. Let F denote the filtration of N, induced by F_2. The assumption is easily seen to imply that

$$F^{s_0} N = F^s N \qquad \text{for } s \geq s_0.$$

The filtration F of N is $\mathfrak{I}$-stable and therefore separated. $\qquad\qquad \square$

10.4   A projection formula

In this section $f : A \to B$ denotes an epimorphism of noetherian local rings, and M denotes an A-module having a finite free resolution, and N denotes a B-module having a finite free resolution. Let $\mathfrak{n}$ denote the maximal ideal in B.

Proposition 10.16. *If $M \otimes_A N$ has finite length, then*

$$\chi_A(M,N) = \sum (-1)^i \chi_B(\text{Tor}_i^A(M,B), N).$$

Proof. Let $L. \to M$ be a finite resolution by free A-modules, and $F. \to N$ a finite resolution by free B-modules. We have a quasi-isomorphism

$$L. \otimes_A N \to L. \otimes_A F.$$

and a canonical isomorphism

$$L. \otimes_A F. \simeq (L. \otimes_A B) \otimes_B F..$$

Conclusion by the following lemma. $\qquad\square$

Lemma 10.17. *Let $F.$ denote a bounded complex of finitely generated free modules and $N.$ a bounded complex with finitely generated homology modules. If $\mathrm{Supp}(F.) \cap \mathrm{Supp}(N.) = \{\mathfrak{n}\}$, then*

$$\chi(F. \otimes_A N.) = \sum(-1)^i \chi(F. \otimes_A H_i(N.)).$$

Proof. This will be done by induction on the length of N. Suppose $N_i = 0$ for $i > 0$. We have a short exact sequence of complexes

$$0 \longrightarrow N'. \longrightarrow N. \longrightarrow H_0(N.) \longrightarrow 0$$

where $N'.$ is the complex

$$\cdots \longrightarrow N_2 \longrightarrow N_1 \longrightarrow \mathrm{Im}(\partial_1) \longrightarrow 0.$$

The additivity of the Euler characteristic gives

$$\chi(F. \otimes_A N.) = \chi(F. \otimes_A N'.) + \chi(F. \otimes_A H_0(N.)).$$

On the other hand, we have a quasi-isomorphism $N''. \to N'.$, where $N''.$ is the complex

$$\cdots \longrightarrow N_2 \longrightarrow \mathrm{Ker}(\partial_1) \longrightarrow 0 \longrightarrow 0.$$

From this we deduce a quasi-isomorphism $F. \otimes_A N''. \to F. \otimes_A N'.$ and we get $\chi(F. \otimes_A N'.) = \chi(F. \otimes_A N''.)$. We leave the final details to the reader. $\quad\square$

10.5 Power series over a field

In this section k denotes a field and $A = k[[X_1, \ldots, X_d]]$ the power series ring in d variables. The maximal ideal in A is denoted $\mathfrak{m}$.

Definition 10.18. If M and N are A-modules, we define the *completed tensor product*.

$$M \hat{\otimes}_k N = \varprojlim (M/\mathfrak{m}^i \otimes_k N/\mathfrak{m}^i N).$$

We have clearly constructed an additive functor in two variables from the category of A-modules to that of $A \hat{\otimes}_k A$-modules. Note

$$A \hat{\otimes}_k A \simeq k[[T_1, \ldots, T_d, S_1, \ldots, S_d]]$$

and note that $(a, b) \to ab$ induces a canonical map $A \hat{\otimes}_A A \to A$, the *diagonal*.

Proposition 10.19. *Let M be a finitely generated A-module. The functor*

$$N \mapsto N \hat{\otimes}_k M$$

is exact on the category for finitely generated modules.

Proof. Given an exact sequence

$$0 \longrightarrow P \longrightarrow N \longrightarrow Q \longrightarrow 0$$

we deduce two exact sequences

$$0 \longrightarrow P/P \cap \mathfrak{m}^i N \otimes_k M/\mathfrak{m}^i M \longrightarrow N/\mathfrak{m}^i N \otimes_k M/\mathfrak{m}^i M$$
$$\longrightarrow Q/\mathfrak{m}^i Q \otimes_k M/\mathfrak{m}^i M \longrightarrow 0$$

and

$$0 \longrightarrow P/\mathfrak{m}^i P \otimes_k M/\mathfrak{m}^i M \longrightarrow P/P \cap \mathfrak{m}^i N \otimes_k M/\mathfrak{m}^i M$$
$$\longrightarrow P \cap \mathfrak{m}^i N/\mathfrak{m}^i P \otimes_k M/\mathfrak{m}^i M \longrightarrow 0.$$

From the Artin–Rees lemma follows that there exists $r \in \mathbb{N}$ such that

$$P \cap \mathfrak{m}^{i+r} N/\mathfrak{m}^{i+r} P \to P \cap \mathfrak{m}^i N/\mathfrak{m}^i P$$

is zero for $i \gg 0$. It follows that

$$\varprojlim N \cap \mathfrak{m}^i N/\mathfrak{m}^i P = 0.$$

Conclusion by the following lemma. $\square$

Lemma 10.20. *Let*

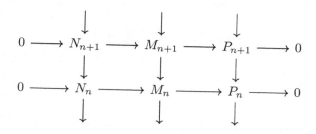

denote a short exact sequence of projective systems indexed by $\mathbb{N}$ of A-modules of finite length. Then

$$0 \longrightarrow \varprojlim N_n \longrightarrow \varprojlim M_n \longrightarrow \varprojlim P_n \longrightarrow 0$$

is exact.

Proof. Fix $n \in \mathbb{N}$ for a moment. The projections $M_{n+i} \to M_n$ have the same image for $i \gg 0$. Let L_n denote that image. It is clear that $M_{n+i} \to M_n$ induces an epimorphism $L_{n+1} \to L_n$. It follows from the first part of 4.3 that $\varprojlim N_n \to \varprojlim N_n/L_n$ is an epimorphism. Consider now the exact sequence of projective systems

$$0 \longrightarrow M_n/L_n \longrightarrow N_n/L_n \longrightarrow P_n \longrightarrow 0.$$

It follows from the second part of lemma 4.3 that $\varprojlim N_n/L_n \simeq \varprojlim P_n$. $\square$

Proposition 10.21. *Let M and N be finitely generated A-modules, we have a canonical isomorphism*

$$(M \hat{\otimes}_k N) \otimes_{A \hat{\otimes}_k A} A \simeq M \otimes_A N.$$

Proof. We have a natural map from left to right induced by

$$M/\mathfrak{m}^i M \otimes_k N/\mathfrak{m}^i N \to M/\mathfrak{m}^i M \otimes_A N/\mathfrak{m}^i N$$

and the canonical isomorphism

$$M \otimes_A N \simeq \varprojlim(M/\mathfrak{m}^i M \otimes_A N/\mathfrak{m}^i N).$$

From the fact that $\hat{\otimes}_k$ is right exact follows that it suffices to check the case $M = N = A$, which we leave to the reader. $\square$

Proposition 10.22. *Let M and N be finitely generated A-modules such that $M \otimes_A N$ has finite length. Then $(M \hat{\otimes}_k N) \otimes_{A \hat{\otimes}_k A} A$ has finite length and*

$$\chi_A(M, N) = \chi_{A \hat{\otimes}_k A}(M \hat{\otimes}_k N, A).$$

Proof. The first part follows immediately from 10.21. Let now $L.$ and $F.$ be finite free resolutions. From exactness of $\hat{\otimes}_k$ follows that $L. \hat{\otimes}_k F.$ is a free resolution of $M \hat{\otimes}_k N$. Whence

$$\operatorname{Tor}_{\cdot}^{A \hat{\otimes}_k A}(M \hat{\otimes}_k N, A) = H_{\cdot}((L. \hat{\otimes}_k F.) \otimes_{A \hat{\otimes}_k A} A).$$

By 10.21 we have

$$(L. \hat{\otimes}_k F.) \otimes_{A \hat{\otimes}_k A} A \simeq L. \otimes_A F..$$

By taking homology we get

$$\operatorname{Tor}_{\cdot}^{A \hat{\otimes}_k A}(M \hat{\otimes}_k N, A) = \operatorname{Tor}_{\cdot}^{A}(M, N). \qquad \square$$

Proposition 10.23. *Let $M \neq 0$ and $N \neq 0$ be finitely generated A-modules. Then*

$$\dim(M \hat{\otimes}_k N) = \dim(M) + \dim(N).$$

Proof. Let $\mathfrak{n}$ denote the maximal ideal of $A \hat{\otimes}_k A$. We leave it to the reader to construct a $\mathfrak{n}$-stable filtration F of $M \hat{\otimes}_k N$ such that

$$M \hat{\otimes}_k N / F^i(M \hat{\otimes}_k N) = M/\mathfrak{m}^i M \otimes_k N/\mathfrak{m}^i N.$$

From this follows

$$\chi_F(M \hat{\otimes}_k N, t) = \chi_{\mathfrak{m}}(M, t)\chi_{\mathfrak{m}}(N, t). \qquad \square$$

Proof of 10.1 and 10.5 in case $R = k$ a field. The diagonal $A \hat{\otimes}_k A \to A$ can be described

$$k[[T_1, \ldots, T_d, S_1, \ldots, S_d]] \to k[[X_1, \ldots, X_d]],$$

where $T_i, S_i \to X_i$. It follows that the Koszul complex

$$K.(T_1 - S_1, \ldots, T_d - S_d)$$

will give a resolution for the $A \hat{\otimes}_k A$-module A. We can now apply 10.12 to the $A \hat{\otimes}_k A$-module $M \hat{\otimes}_k N$. $\qquad \square$

10.6 Power series over a discrete valuation ring

In this section R denotes a complete discrete valuation ring and $A = R[[X_1, \ldots, X_d]]$ the power series ring i d variables. The maximal ideal of A will be denoted $\mathfrak{m}$.

Definition 10.24. If M and N are A-modules, we have the completed tensor product

$$M \mathbin{\hat{\otimes}}_R N = \varprojlim (M/\mathfrak{m}^i M \otimes_R M/\mathfrak{m}^i M).$$

We have clearly constructed an additive functor in two variables from the category of A-modules to that of $A \mathbin{\hat{\otimes}}_R A$-modules. Note

$$A \mathbin{\hat{\otimes}}_R A \simeq R[[T_1, \ldots, T_d, S_1, \ldots, S_d]].$$

Proposition 10.25. *Let M be a finitely generated A-module. The functor*

$$N \to N \mathbin{\hat{\otimes}}_R M$$

is right exact on the category for finitely generated modules. If M is R-flat, then the functor is exact.

Proof. By a module we understand a finitely generated A-module. For modules M and N we put

$$\widehat{\mathrm{Tor}}_1(M, N) = \varprojlim \mathrm{Tor}_1^R(M/\mathfrak{m}^i M, N/\mathfrak{m}^i N).$$

An exact sequence of modules

$$0 \longrightarrow P \longrightarrow N \longrightarrow Q \longrightarrow 0$$

gives rise to a six term exact sequence

$$0 \longrightarrow \widehat{\mathrm{Tor}}_1(M, P) \longrightarrow \widehat{\mathrm{Tor}}_1(M, N) \longrightarrow \widehat{\mathrm{Tor}}_1(M, N)$$
$$\longrightarrow M \mathbin{\hat{\otimes}}_R P \longrightarrow M \mathbin{\hat{\otimes}}_R N \longrightarrow M \mathbin{\hat{\otimes}}_R Q \longrightarrow 0$$

as it follows by the proof of 10.19. Suppose now that M is R torsion free. Let π be a local parameter for R. The exact sequence

$$0 \longrightarrow M \overset{\pi}{\longrightarrow} M \longrightarrow M/\pi M \longrightarrow 0$$

gives rise to an injection

$$0 \longrightarrow \widehat{\mathrm{Tor}}_1(M, N) \overset{\pi}{\longrightarrow} \widehat{\mathrm{Tor}}_1(M, N).$$

Thus $\widehat{\mathrm{Tor}}_1(M, N) = 0$, whenever N is annihilated by a power of π. But quite generally

$$\widehat{\mathrm{Tor}}_1(M, N) = \varprojlim \widehat{\mathrm{Tor}}_1(M, N/\mathfrak{m}^i N)$$

as one sees as follows. $\widehat{\mathrm{Tor}}_1(M, N)$ can be calculated as the limit of the projective system $\mathrm{Tor}_1^R(M/\mathfrak{m}^p M, N/\mathfrak{m}^q)$ as $(p, q) \in \mathbb{N} \times \mathbb{N}$, since the diagonal in $\mathbb{N} \times \mathbb{N}$ form a cofinal set.

$$\varprojlim_{p,q} \mathrm{Tor}_1^R(M/\mathfrak{m}^p M, N/\mathfrak{m}^q N) = \varprojlim_q \varprojlim_p \mathrm{Tor}_1^R(M/\mathfrak{m}^p M, N/\mathfrak{m}^q N)$$

but clearly

$$\varprojlim_p \mathrm{Tor}_1^R(M/\mathfrak{m}^p M, N/\mathfrak{m}^q N) = \widehat{\mathrm{Tor}}_1(M, N/\mathfrak{m}^q N). \qquad \square$$

Corollary 10.26. *Let $F.$ be a bounded above complex of finitely generated A-modules, which are R-flat. Then*

(1) *For any quasi-isomorphism $f. : X. \to Y.$ of finitely generated A-modules, $f. \hat{\otimes} 1 : X. \hat{\otimes}_R F. \to Y. \hat{\otimes}_R F.$ is a quasi-isomorphism.*

(2) *If $H.(F.) = 0$ then for a finitely generated A-module M, we have $H.(M \hat{\otimes}_R F.) = 0$.*

Proof. To prove the first part proceed as in 7.37. For finitely generated A-modules M, N choose free resolutions $F. \to M$ and $L. \to N$ and define

$$\widehat{\mathrm{Tor}}.(M, N) = H.(F. \hat{\otimes}_R L.).$$

It follows from the first part, that

$$\widehat{\mathrm{Tor}}.(M, N) = H.(M \hat{\otimes}_R L.) = H.(F. \hat{\otimes}_R N).$$

In particular $\widehat{\mathrm{Tor}}_i(M, N) = 0$, $i > 0$ if M is flat as R-module.

To prove the last part consider a short exact sequence

$$0 \longrightarrow F_2 \longrightarrow F_1 \longrightarrow F_0 \longrightarrow 0$$

of finitely generated A-modules, where F_1, F_0 are R-flat. It follows that $F.$ is R-flat. By our previous result $\widehat{\mathrm{Tor}}_1(F_i, N) = 0$, whence the sequence

$$0 \longrightarrow F_2 \hat{\otimes}_R N \longrightarrow F_1 \hat{\otimes}_R N \longrightarrow F_0 \hat{\otimes}_R N \longrightarrow 0$$

is exact. In general split the complex $F.$ into short exact sequences, tensor with N and fit the resulting short exact sequences together again. $\qquad \square$

Proposition 10.27. *Let M and N be finitely generated A-modules, we have a canonical isomorphism*

$$(M \mathbin{\hat{\otimes}_k} N) \otimes_{A \hat{\otimes}_R A} A \simeq M \otimes_A N.$$

Proof. The product $(a, b) \to ab$ induces a canonical map $A \mathbin{\hat{\otimes}_A} A \to A$. Conclusion by right exactness of $\hat{\otimes}$. $\qquad\square$

Proposition 10.28. *Let M and N be finitely generated A-modules such that $M \otimes_A N$ has finite length. If M is R-flat, then*

$$\chi_A(M, N) = \chi_{A \hat{\otimes}_R A}(M \mathbin{\hat{\otimes}_R} N, A).$$

Proof. Let now $L. \to M$ and $F. \to N$ be finite free resolutions. We have quasi-isomorphisms

$$L. \mathbin{\hat{\otimes}_R} F. \longrightarrow L. \mathbin{\hat{\otimes}_R} N \longrightarrow M \mathbin{\hat{\otimes}_R} N$$

as it follows from 10.26. This proves that $L. \mathbin{\hat{\otimes}_R} F.$ is a free resolution of $M \mathbin{\hat{\otimes}_R} N$. As in 10.22 we conclude

$$\mathrm{Tor}_{\cdot}^{A \hat{\otimes}_R A}(M \mathbin{\hat{\otimes}_R} N, A) = \mathrm{Tor}_{\cdot}^{A}(M, N). \qquad\square$$

Proposition 10.29. *Let $M \neq 0$ and $N \neq 0$ be finitely generated A-modules which are both R-flat. Then*

$$\dim(M \mathbin{\hat{\otimes}_R} N) = \dim(M) + \dim(N) - 1.$$

Proof. Let $\pi \in R$ be a local parameter and $k = R/(\pi)$. The short exact sequence

$$0 \longrightarrow M \xrightarrow{\ \pi\ } M \longrightarrow M/\pi M \longrightarrow 0$$

gives rise to the exact sequence

$$0 \longrightarrow M \mathbin{\hat{\otimes}_R} N \xrightarrow{\ \pi\ } M \mathbin{\hat{\otimes}_R} N \longrightarrow M/\pi M \mathbin{\hat{\otimes}_R} N \longrightarrow 0.$$

From this follows that π is a nonzero divisor on $M \mathbin{\hat{\otimes}_R} N$ and that

$$M \mathbin{\hat{\otimes}_A} N/\pi(M \mathbin{\hat{\otimes}_R} N) \simeq (M/\pi M) \mathbin{\hat{\otimes}_R} (N/\pi N).$$

Conclusion by 10.23 $\qquad\square$

Proof of 10.1 and 10.5 in case R is a complete discrete valuation ring.
It suffices to treat the case where M is either R-flat or annihilated by π a local parameter for R, as it follows from 10.6.

Case (1). π is a nonzero divisor for M and N. This means that M, N are R-flat, and the method of section 10.5 above can be used.

Case (2). π is a nonzero divisor for M and annihilates N. Put $B = A/(\pi)$. By 10.16 we have

$$\chi_A(M, N) = \chi_B(\mathrm{Tor}_0^A(M, B), N) - \chi_B(\mathrm{Tor}_1^A(M, B), N).$$

Since π is a nonzero divisor on M, we get

$$\chi_A(M, N) = \chi_B(M/\pi M, N).$$

It is now easy to conclude from the fact that Serre's conjecture 10.1, 10.3 and 10.4 are true for B, see the end of 10.5.

Case (3). π annihilates M, N. From the formula above we get

$$\chi_A(M, N) = \chi_B(M, N) - \chi_B(M, N) = 0$$

and

$$\dim(M) + \dim(N) \le \dim(B) < \dim(A)$$

since 10.1 is true for B. □

10.7 Application of Cohen's structure theorem

Let A denote a complete noetherian local ring with maximal ideal $\mathfrak{m}$. The following theorem is due to I.S. Cohen.

Theorem 10.30. *Let $p = \mathrm{char}(A/\mathfrak{m})$.*

(1) *If $p = 0$, then there exists a subfield k of A such that $k \to A/\mathfrak{m}$ is an isomorphism.*

(2) *If $p > 0$, then there exists a complete discrete valuation ring R whose maximal ideal is generated by p and a morphism of local rings $R \to A$ such that $R/(p) \to A/\mathfrak{m}$ is an isomorphism.*

Proof. The proof will not be given here, see Grothendieck, Eléments de géométrie algébrique, Inst. Hautes Études Sci. Pub. Math. 20 (1964), section IV.0.19.8, or Nagata, Local rings, Interscience Publ. 1962, section 31.

Let us indicate the principle for the proof of (2). One proves first, that a complete discrete valuation ring R whose maximal ideal is generated by p has the following lifting property by Hensel:

> *For any complete noetherian local ring C and $\mathfrak{I}$ an ideal in C, any*
> *local morphism $R \to C/\mathfrak{I}$ can be lifted to a local morphism $R \to C$.*

Second, one proves that given a field k of characteristic $p > 0$ then there exists R as above with residue field isomorphic to k. (In case k is perfect one can use the ring of infinite Witt vectors). □

Corollary 10.31. *The complete local ring A is isomorphic to a quotient ring of $k[[T_1, \ldots, T_d]]$ or $R[[T_1, \ldots, T_d]]$, where k or R are as above.*

Proof. Let $a_1, \ldots, a_n$ generate the maximal ideal of A. In case $p = 0$ we have a map

$$k[[T_1, \ldots, T_d]] \to A, \quad T_i \to a_i$$

which is surjective according to 5.11.

In case $p > 0$ we have a map

$$R[[T_1, \ldots, T_d]] \to A, \quad T_i \to a_i$$

which is surjective again according to 5.11. □

A local ring is called *equicharacteristic* if it contains a field.

Corollary 10.32. *An equicharacteristic complete regular local ring is isomorphic to a ring of the form*

$$k[[T_1, \ldots, T_n]]$$

where k is a field.

Proof. Let $a_1, \ldots, a_n$ generate the maximal ideal of A, with $\dim(A) = n$. As in the proof of 10.31 we get an epimorphism

$$k[[T_1, \ldots, T_n]] \to A, \quad T_i \to a_i,$$

where $k = R/(p)$ if $p > 0$. For dimension reasons, this is an isomorphism. □

Corollary 10.33. *Any complete regular local ring which is not equicharacteristic ("mixed characteristic") is isomorphic to a ring of the form*

$$R[[T_1, \ldots, T_n]]/(a)$$

where a is a power series whose constant term belongs to $\mathfrak{n}$ but not to $\mathfrak{n}^2$, $\mathfrak{n}$ being the maximal ideal in R.

Proof. If $\dim(A) = n$ we obtain an epimorphism

$$R[[T_1, \ldots, T_n]] \to A.$$

The kernel $\mathfrak{p}$ is a prime ideal with $\dim(R[[T_1, \ldots, T_n]]/\mathfrak{p}) = n$, whence $\mathfrak{p}$ is principal, say, $\mathfrak{p} = (a)$. The maximal ideal of $R[[T_1, \ldots, T_n]]$ is the set of power series whose constant term belongs to $\mathfrak{n}$. $\qquad\square$

Corollary 10.34. *Let A be a complete regular local ring such that $p \in \mathfrak{m} - \mathfrak{m}^2$. Then A is isomorphic to a ring of the form*

$$R[[T_1, \ldots, T_{n-1}]].$$

Proof. Let $\dim(A) = n$, we can find $a_1, \ldots, a_{n-1} \in \mathfrak{m}$ such that $\mathfrak{m}$ is generated by $a_1, \ldots, a_{n-1}, p$. The morphism

$$R[[T_1, \ldots, T_{n-1}]] \to A, \quad T_i \to a_i$$

is an epimorphism by 5.11. For dimension reasons this is an isomorphism. $\qquad\square$

Proof of 10.1 in the general case. We can assume that A is complete, 10.7.

The theorem is already proved in case A is equicharacteristic. In the mixed characteristic case we have with the notation of 10.33, $A \simeq R[[T_1, \ldots, T_n]]/(a)$. Let now M, N be finitely generated A-modules such that $M \otimes_A N$ has finite length. Consider M as a $C = R[[T_1, \ldots, T_n]]$-module. We have by 10.16

$$\chi_C(M, N) = \sum (-1)^i \chi_A(\mathrm{Tor}_i^C(M, A), N).$$

Let us calculate $\mathrm{Tor}^C(M, A)$. We have a free resolution

$$0 \longrightarrow C \overset{a}{\longrightarrow} C \longrightarrow A \longrightarrow 0,$$

from which we get that $\mathrm{Tor}_i^C(M, A) = 0$, $i \neq 0, 1$ and $\mathrm{Tor}_0^C(M, A) = \mathrm{Tor}_1^C(M, A) = M$, whence $\chi_C(M, N) = 0$ and therefore by 10.5

$$\dim(M) + \dim(N) < \dim(C)$$

and consequently

$$\dim(M) + \dim(N) \leq \dim(A). \qquad\square$$

10.8 The amplitude inequality

In this section we shall draw various consequences of the dimension inequality 10.1. Throughout this section A denotes a regular local ring.

Theorem 10.35. *Let $M \neq 0$ and $N \neq 0$ be finitely generated A-modules. Then*

$$\dim(M) + \dim(N) \leq \dim(M \otimes_A N) + \dim(A).$$

Proof. Let $\mathfrak{p} \in \operatorname{Supp}(M)$ be such that $\dim(M) = \dim(A/\mathfrak{p})$ and $\mathfrak{q} \in \operatorname{Supp}(N)$ be such that $\dim(N) = \dim(A/\mathfrak{q})$. Let $\mathfrak{r}$ be minimal in $A/\mathfrak{p} + \mathfrak{q}$. We have by 10.1

$$\dim((A/\mathfrak{p})_{\mathfrak{r}}) + \dim((A/\mathfrak{q})_{\mathfrak{r}}) \leq \dim(A_{\mathfrak{r}}).$$

Add $2\dim(A/\mathfrak{r})$ to both sides to get

$$\dim(A/\mathfrak{p}) + \dim(A/\mathfrak{q}) \leq \dim(A/\mathfrak{r}) + \dim(A)$$

since $\mathfrak{r} \in \operatorname{Supp}(M \otimes_A N)$, we have

$$\dim(A/\mathfrak{r}) \leq \dim(M \otimes_A N). \qquad \square$$

Corollary 10.36. *Let $L.$ and $F.$ be bounded complexes of finitely generated free A-modules. If $H.(L.) \neq 0$ and $H.(F.) \neq 0$ then*

$$\dim(L. \otimes_A F.) \geq \dim(L.) + \dim(F.) - \dim(A).$$

Proof. We can assume $L_i = F_i = 0$ for $i < 0$, $H_0(L.) \neq 0$ and $H_0(F.) \neq 0$. From 9.22 and 9.27 we get

$$\begin{aligned}
\dim(L. \otimes_A F.) &= \sup_{r,s}(\dim(H_r(L.) \otimes_A H_s(F.)) - r - s) \\
&\geq \sup_{r,s}(\dim(H_r(L.)) + \dim(H_s(F.)) - r - s + \dim(A)) \\
&= \dim(L.) + \dim(F.) - \dim(A). \qquad \square
\end{aligned}$$

Corollary 10.37. *Let $L.$ and $F.$ be as in 10.36. Then*

$$\operatorname{amp}(L. \otimes_A F.) \geq \operatorname{amp}(L.) + \operatorname{amp}(F.).$$

Proof. The complex A is dualizing. We have

$$\dim(\mathrm{Hom}_A^{\bullet}(L_{\bullet}, A) \otimes_A \mathrm{Hom}_A^{\bullet}(F_{\bullet}, A))$$
$$\geq \dim(\mathrm{Hom}_A^{\bullet}(L_{\bullet}, A) + \dim \mathrm{Hom}_A^{\bullet}(F_{\bullet}, A)) - \dim(A).$$

On the other hand, by 9.23

$$\dim(\mathrm{Hom}_A^{\bullet}(L_{\bullet}, A)) = \mathrm{depth}(L_{\bullet}) + \mathrm{amp}(L_{\bullet})$$
$$\dim(\mathrm{Hom}_A^{\bullet}(F_{\bullet}, A)) = \mathrm{depth}(F_{\bullet}) + \mathrm{amp}(F_{\bullet})$$
$$\dim(\mathrm{Hom}_A^{\bullet}(L_{\bullet}, A) \otimes_A \mathrm{Hom}_A^{\bullet}(F_{\bullet}, A))$$
$$= \mathrm{depth}(L_{\bullet} \otimes_A F_{\bullet}) + \mathrm{amp}(L_{\bullet} \otimes_A F_{\bullet})$$

finally by 9.4 and 9.5

$$\mathrm{depth}(L_{\bullet} \otimes_A F_{\bullet}) + \dim(A) = \mathrm{depth}(L_{\bullet}) + \mathrm{depth}(F_{\bullet}). \qquad \square$$

10.9   Translation invariant operators

In this section we study a class of linear operators on the polynomial ring $\mathbb{Q}[t]$. For $a \in \mathbb{Q}$ let translation by a be denoted

$$\Lambda_a : \mathbb{Q}[t] \to \mathbb{Q}[t], \quad P(t) \mapsto P(t + a)$$

clearly

$$\Lambda_a \Lambda_b = \Lambda_{a+b}, \qquad a, b \in \mathbb{Q}.$$

Definition 10.38. A linear map $\psi : \mathbb{Q}[t] \to \mathbb{Q}[t]$ is called a translation invariant operator if

$$\psi \Lambda_a = \Lambda_a \psi \qquad \text{for all } a \in \mathbb{Q}.$$

A typical example is D, $P(t) \mapsto P'(t)$ derivation with respect to t and the operator Δ, $P(t) \mapsto P(t + 1) - P(t)$ introduced in 1.26.

Proposition 10.39. *Any translation invariant operator* $\psi : \mathbb{Q}[t] \to \mathbb{Q}[t]$ *can be written uniquely*

$$\psi = \sum_{k=0}^{\infty} b_k D^k, \qquad b_k \in \mathbb{Q}.$$

Proof. Let $a \in \mathbb{Q}$ and $P(t) \in \mathbb{Q}[[t]]$. By Taylor's formula we have

$$\Lambda_a(P(t)) = P(t + a) = \sum_k \frac{1}{k!} P^{(k)}(a) t^k.$$

Apply ψ to this identity to get

$$\Lambda_a(\psi(P(t))) = \sum_k \frac{1}{k!} P^{(k)}(a) \psi(t^k).$$

Substitute $t = 0$ in this formula to get

$$\psi(P(a)) = \sum_k \frac{1}{k!} \psi(t^k)_{|t=0} D^k P(a).$$

This shows that the two polynomials above take the same value for all $a \in \mathbb{Q}$. Thus they must be identical.

To prove uniqueness, note that

$$\psi(t^k)\big|_{t=0} = k! b_k. \tag{10.1}$$

$\square$

This shows that the ring of translation invariant operators is isomorphic to the ring of power series $\mathbb{Q}[[D]]$, in particular it is commutative.

Definition 10.40. Consider a translation invariant operator $\psi \neq 0$,

$$\psi = \sum_{k=0}^{\infty} b_k D^k, \qquad b_k \in \mathbb{Q}.$$

The smallest number $k \in \mathbb{N}$ such that $b_k \neq 0$ is called the *order* of ψ and is denoted $\nu(\psi)$. Define the *multiplicity* of ψ by

$$e(\psi) = b_k, \qquad k = \nu(\psi).$$

Remark 10.41. Let $\psi \neq 0$ be a translation invariant operator of order p. Then $\psi : \mathbb{Q}[t] \to \mathbb{Q}[t]$ is surjective, its kernel is the space of polynomials of degree less that p. If $P(t) \neq 0$ is a polynomial of degree p, we have

$$\deg(\psi(P(t))) = \deg(P(t)) - \nu(\psi).$$

Moreover

$$\psi\left(\frac{t^p}{p!}\right) = e(\psi), \qquad p = \nu(\psi).$$

Remark 10.42. Let $\psi \neq 0$ be a translation invariant operator which transforms the set of $P(t) \in \mathbb{Q}[t]$ for which $P(d) \in \mathbb{Z}$ for large values of $d \in \mathbb{N}$, into itself. Then $e(\psi) \in \mathbb{Z}$, as it follows from 10.41.

Lemma 10.43. *The operators $(\Lambda_a)_{a \in \mathbb{Q}}$ are linearly independent over $\mathbb{Q}$.*

Proof. For $a \in \mathbb{Q}$ we have

$$\Lambda_a = \sum \frac{a^k}{k!} D^k.$$

Thus it will suffice to prove that if $a_1, \ldots, a_n \in \mathbb{Q}$ are distinct then

$$\exp(a_1 D), \ldots, \exp(a_n D)$$

are linearly independent elements in $\mathbb{Q}[[D]]$. Given a linear combination

$$\lambda_1 \exp(a_1 D) + \cdots + \lambda_n \exp(a_n D) = 0$$

differentiate this $(n-1)$-times and use that the Vandermonde determinant with respect to $a_1, \ldots, a_n$ is different from zero. □

10.10   Todd operators

In this section A denotes a graded ring 1.6. A graded module which can be decomposed into a finite direct sum of graded modules of the form $A[n]$, $n \in \mathbb{Z}$, is called a finitely generated *free graded* module. In the rest of this section we assume $A \neq 0$.

Lemma 10.44. *Let $n_1, \ldots, n_s$ and $m_1, \ldots, m_r$ be two decreasing sequences of integers. If $\bigoplus_{j=1}^{s} A[n_j] \simeq \bigoplus_{i=1}^{r} A[m_i]$ then the two sequences are identical.*

Proof. Let L be a free graded module isomorphic to $\bigoplus_{j=1}^{s} A[n_j]$. Let p be the smallest integer for which $[L]_p \neq 0$. We have $n_1 = \cdots = n_p = -p$ and $n_{p+1} < n_p$. Moreover

$$L/A[L]_p \simeq \bigoplus_{i=2}^{r} A[n_i] \simeq \bigoplus_{i=2}^{r} A[m_i].$$

Conclusion by induction on r. □

Definition 10.45. Let $L.$ be a bounded complex of finitely generated free graded modules. For $i \in \mathbb{Z}$, let $L_i \simeq \bigoplus_j A[n_{ij}]$. Define the *Todd operator* $\chi_L.$ to be (with notation of section 10.9)

$$\chi_L. = \sum_{i,j} (-1)^i \Lambda_{n_{ij}}.$$

The following properties are easily verified

$$\chi_{L.\oplus F.} = \chi_L. + \chi_F.$$
$$\chi_L. = 0 \quad \text{if } H.(L.) = 0$$
$$\chi_{L.\otimes_A F.} = \chi_L. \chi_F.$$

Example 10.46. The Koszul complex in 10.12 gives rise to an operator of order r and multiplicity 1.

The complex considered in 10.14 gives an operator of order 2 and multiplicity $\binom{k+1}{2}$.

The reader is invited to discuss the Gulliksen–Negaard complex 2.66 along the same lines and find an operator of order 4 and multiplicity $\frac{n}{2}\binom{n+1}{3}$.

These examples all illustrate the following general proposition.

Proposition 10.47. *Given a complex $L.$ as above*

$$0 \longrightarrow L_s \longrightarrow L_{s-1} \longrightarrow \cdots \longrightarrow L_1 \longrightarrow L_0 \longrightarrow 0$$

and let $L_i \simeq \bigoplus_j A[-n_{ij}]$. Suppose the following two conditions are satisfied

(1) $n_{ij} \geq n_{i-1,j'}$ *for all i, j, j'.*

(2) *There exists $j_s, j_{s-1}, \ldots, j_0$ such that $n_{sj_s} > n_{s-1,j_{s-1}} > \cdots > n_{0j_0}$.*

Then $\chi_L. \neq 0$ and has order $\leq s$. If the order of $\chi_L.$ is s, then the multiplicity of $\chi_L.$ is positive.

Proof. We have (10.1)

$$\chi_L. = \sum (-1)^k \frac{1}{k!} \rho_k D^k$$

where

$$\rho_k = \sum_i (-1)^i \sum_j n_{ij}^k.$$

Suppose $\rho_0 = \rho_1 = \cdots = \rho_s = 0$ and put $r_i = \operatorname{rank}_A(L_i)$. The following determinant is zero as the alternating sum of its column vectors is zero,

$$
E = \begin{vmatrix}
r_0 & \cdots & r_s \\
\sum_j n_{0j} & \cdots & \sum_j n_{sj} \\
& \cdots & \\
\sum_j n_{0j}^s & \cdots & \sum_j n_{sj}^s
\end{vmatrix}
$$

and therefore the following sum of determinants is zero

$$
\sum_{j_0,\ldots,j_s} \begin{vmatrix}
1 & \cdots & 1 \\
n_{0j} & \cdots & n_{sj} \\
& \cdots & \\
n_{0j}^s & \cdots & n_{sj}^s
\end{vmatrix}.
$$

The determinant corresponding to $j_0, \ldots, j_s$ equals

$$
\prod_{p>q} (n_{p,j_p} - n_{q,j_q})
$$

according to Vandermonde. The first assumption above assures that all of these factors are positive and the second that at least one is strictly positive.

Suppose the order of $\chi_{L.}$ equals s. Replace the first column in the determinant E above by the alternating sum of all columns to get

$$
E = \begin{vmatrix}
0 & r_1 & \cdots & r_s \\
0 & \sum_j n_{1j} & \cdots & \sum_j n_{sj} \\
& & \cdots & \\
\rho_s & \sum_j n_{1j}^s & \cdots & \sum_j n_{sj}^s
\end{vmatrix}
$$

and whence $E = (-1)^s \rho_s E'$ there E' is the subdeterminant obtained from E by deleting the first column and the last row. E' is positive by the same argument which proved E positive. $\qquad\square$

Example 10.48. Consider a complex $L.$ of the form

$$
0 \longrightarrow A[-3] \longrightarrow A[-1] \oplus A[-1] \longrightarrow A \longrightarrow 0.
$$

The Todd operator $\chi_{L.}$ has order 1 and multiplicity -1.

10.11 Serre's conjecture in the graded case

Let A denote a graded noetherian ring such that A_+ is generated by A_1 and A_0 is a Artin local ring. Let $\mathfrak{v}$ denote the *vertex* of A, i.e., the maximal ideal of A containing A_+. Note that the elements of $A - \mathfrak{v}$ are nonzero divisors on any graded A-module different from zero.

For a finitely generated graded A-module M, let $\chi(t, M)$ denote the Samuel polynomial, $\chi(t, M) \in \mathbb{Q}[t]$, and

$$\chi(n, M) = \sum_{i<n} \ell_{A_0}(M_i), \qquad n \gg 0.$$

Proposition 10.49. *With the notation above, let $M \neq 0$ be a finitely generated graded A-module, then*

$$\dim(M) = \deg(\chi(t, M)).$$

Proof. The filtration F of M given by $F^t M = \bigoplus_{i \geq t} M_i$ is easily seen to be A_+-stable, since A_+ is generated by A_1. Moreover

$$M_\mathfrak{v}/F^t M_\mathfrak{v} = M/F^t M.$$

The result now follows from the proof of 1.42. $\qquad\square$

Definition 10.50. Let $M \neq 0$ be a finitely generated graded A-module of dimension d. Write

$$\chi(t, M) = e(M)\binom{t}{d} + \text{lower terms}.$$

$e(M)$ is a strictly positive integer, which we call the *multiplicity* of M, 1.27.

Proposition 10.51. *Suppose $M \neq 0$ has a finite resolution by finitely generated free graded modules $L..$. Then for any finitely generated graded module N,*

$$\sum_i (-1)^i \chi(t, H_i(L. \otimes_A N)) = \chi_{L.}(\chi(t, N))$$

where $\chi_{L.}$ is the Todd operator of $L..$.

Proof. Straightforward. $\qquad\square$

Corollary 10.52. *With the notation above, the Todd operator $\chi_{L.}$ has order $\dim(A_\mathfrak{v}) - \dim(M_\mathfrak{v})$ and multiplicity $e(M)/e(A)$.*

Proof. Take $A = N$ in 10.51 and apply the results of section 10.9. $\qquad\square$

Corollary 10.53. *Let $N \neq 0$ be a finitely generated graded module such that $M \otimes_A N$ has finite length, then*

(1) $\dim(M_\mathfrak{v}) + \dim(N_\mathfrak{v}) \leq \dim(A_\mathfrak{v})$.

(2) $\dim(M_\mathfrak{v}) + \dim(N_\mathfrak{v}) = \dim(A_\mathfrak{v}) \Rightarrow \chi_{A_\mathfrak{v}}(M_\mathfrak{v}, N_\mathfrak{v}) = e(M)e(N)/e(A)$.

(3) $\dim(M_\mathfrak{v}) + \dim(N_\mathfrak{v}) < \dim(A_\mathfrak{v}) \Rightarrow \chi_{A_\mathfrak{v}}(M_\mathfrak{v}, N_\mathfrak{v}) = 0$.

Proof. We have

$$\dim(M \otimes_A N) \geq \deg(\sum(-1)^i \chi(t, H_i(L. \otimes_A N)))$$

since $\dim(M \otimes_A N) = 0$, $\chi_{L.}(\chi(t, N))$ is a constant polynomial different from zero by 10.51. The first case corresponds to

$$\dim(N_\mathfrak{v}) < \nu(\chi_{L.}) = \dim(A_\mathfrak{v}) - \dim(M_\mathfrak{v}).$$

The second case to

$$\dim(M_\mathfrak{v}) + \dim(N_\mathfrak{v}) = \dim(A_\mathfrak{v}).$$

In that case each of the polynomials $\chi(t, H_i(L. \otimes_A N))$ are constant, so

$$\sum(-1)^i \chi(t, H_i(L. \otimes_A N)) = \chi_{A_\mathfrak{v}}(M_\mathfrak{v}, N_\mathfrak{v}).$$

On the other hand,

$$\chi_{L.}(\chi(t, N)) = e(M)e(A)^{-1}e(N)$$

as it follows from section 10.9. $\qquad\square$

Chapter 11

Complexes of Free Modules

11.1 McCoy's theorem

Throughout this section we let A be a ring, a module is understood as an A-module and a linear map is understood as an A-linear map.

Definition 11.1. Given $f : E \to F$ a linear map of finitely generated free modules. Let X denote a matrix for f with respect to bases of E and F. For $n \in \mathbb{Z}$ let $\mathfrak{I}_n(f)$ denote the ideal generated by n-minors of X. By convention $\mathfrak{I}_n(f) = 0$ if $n > \inf\{\mathrm{rank}_A(E), \mathrm{rank}_A(F)\}$ and $\mathfrak{I}_n(f) = A$ if $n \leq 0$. Note that

$$\cdots \subseteq \mathfrak{I}_2(f) \subseteq \mathfrak{I}_1(f) \subseteq \mathfrak{I}_0(f) = A.$$

Let us recall that for an ideal $\mathfrak{I}$ and a module M, $\Gamma_{\mathfrak{I}}(M) = 0$ is equivalent to "any element of M annihilated by $\mathfrak{I}$ is zero".

Theorem 11.2 (McCoy). *Given a linear map $f : E \to F$ of finitely generated free modules with $\mathrm{rank}_A(E) = n$. For an arbitrary module $M \neq 0$,*

$$f \otimes 1 : E \otimes_A M \to F \otimes_A M$$

is an injection if and only if

$$\Gamma_{\mathfrak{I}_n(f)}(M) = 0.$$

Proof. Let us consider the case where $\mathrm{rank}_A(E) = 3$ and $\mathrm{rank}_A(F) = 4$. Let a matrix for f be given

$$X = \begin{pmatrix} a_{11} & a_{12} & a_{13} \\ a_{21} & a_{22} & a_{23} \\ a_{31} & a_{32} & a_{33} \\ a_{41} & a_{42} & a_{43} \end{pmatrix}.$$

Suppose first that $\mathfrak{I}_3(X)$ does not annihilate any nonzero element of M. Suppose

$$X \begin{pmatrix} m_1 \\ m_2 \\ m_3 \end{pmatrix} = 0$$

for $m_1, m_2, m_3 \in M$. Multiply X from the left by the matrix

$$\left(\begin{vmatrix} a_{22} & a_{23} \\ a_{32} & a_{33} \end{vmatrix}, \quad - \begin{vmatrix} a_{12} & a_{13} \\ a_{32} & a_{33} \end{vmatrix}, \quad \begin{vmatrix} a_{12} & a_{13} \\ a_{22} & a_{23} \end{vmatrix}, \quad 0 \right)$$

to see that

$$\begin{vmatrix} a_{11} & a_{12} & a_{13} \\ a_{21} & a_{22} & a_{23} \\ a_{31} & a_{32} & a_{33} \end{vmatrix}$$

annihilates m_1, etc. Thus $\mathfrak{I}_3(X)$ annihilates m_1, m_2, m_3, so these are zero by assumption. Conversely, suppose $f \otimes 1$ is injective. We are going see, that for $m \in M$ we have

$$\mathfrak{I}_3(X)m = 0 \quad \Rightarrow \quad \mathfrak{I}_2(X)m = 0$$

and

$$\mathfrak{I}_2(X)m = 0 \quad \Rightarrow \quad \mathfrak{I}_1(X)m = 0.$$

This will clearly suffice for the proof. So suppose for example

$$\begin{vmatrix} a_{22} & a_{23} \\ a_{32} & a_{33} \end{vmatrix} m \neq 0$$

then the non trivial vector

$$\left(\begin{vmatrix} a_{22} & a_{23} \\ a_{32} & a_{33} \end{vmatrix} m, \quad - \begin{vmatrix} a_{21} & a_{23} \\ a_{31} & a_{33} \end{vmatrix} m, \quad \begin{vmatrix} a_{21} & a_{22} \\ a_{31} & a_{32} \end{vmatrix} m \right)$$

would be in the kernel of $f \otimes 1$, contradicting injectivity. Next suppose $\mathfrak{I}_2(X)m = 0$ and let us prove $\mathfrak{I}_1(X)m = 0$. Suppose for a moment that $\mathfrak{I}_1(f)m \neq 0$, for example $am \neq 0$. Then the non trivial vector

$$(-bm, am, 0)$$

would be in the kernel of $f \otimes 1$. We trust the reader to prove McCoy's theorem in the general case. $\qquad \square$

Corollary 11.3. *Let $f : E \to E$ be an endomorphism of a finitely generated free module $E \neq 0$. Then f is an injection if and only if $\det(f) \in A$ is a nonzero divisor.*

Proof. Straightforward. □

11.2 The rank of a linear map

We shall keep the conventions from 11.1 and consider a linear map $f : E \to F$ of finitely generated free modules.

Definition 11.4. Given a module $M \neq 0$. Put

$$\text{rank}(f, M) = \sup\{n \in \mathbb{N} \mid \mathfrak{I}_n(f)M \neq 0\},$$
$$\underline{\text{rank}}(f, M) = \sup\{n \in \mathbb{N} \mid \Gamma_{\mathfrak{I}_n(f)}(M) = 0\}.$$

Clearly

$$\underline{\text{rank}}(f, M) \leq \text{rank}(f, M).$$

Remark 11.5. Suppose $M \neq 0$ is such that

$$f \otimes 1 : E \otimes_A M \to F \otimes_A M$$

is an injection. Then by McCoy's theorem

$$\underline{\text{rank}}(f, M) = \text{rank}(f, M) = \text{rank}_A(E).$$

This has the following important generalization.

Theorem 11.6. *Let*

$$E \xrightarrow{\ f\ } F \xrightarrow{\ g\ } G$$

be linear maps between finitely generated free modules with $gf = 0$. Suppose $M \neq 0$ is a module such that

$$E \otimes_A M \xrightarrow{\ f \otimes 1\ } F \otimes_A M \xrightarrow{\ g \otimes_A 1\ } G \otimes_A M$$

is exact. Then

$$\underline{\text{rank}}(f, M) = \text{rank}(f, M) \quad \Rightarrow \quad \underline{\text{rank}}(g, M) = \text{rank}(g, M)$$

and

$$\text{rank}(f, M) + \text{rank}(g, M) = \text{rank}_A(F).$$

The proof will be given towards the end of this section. First some preparation.

Lemma 11.7. *Let I_t denote the identity $t \times t$ matrix. For a matrix of the form*

$$X = \begin{pmatrix} I_t & Z \\ 0 & Y \end{pmatrix}$$

we have for all $n \in \mathbb{N}$

$$\mathfrak{I}_{n+t}(X) = \mathfrak{I}_n(Y).$$

Proof. Elementary column operations will change nothing. By such we can obtain $Z = 0$. The inclusion

$$\mathfrak{I}_n(Y) \subseteq \mathfrak{I}_{n+t}(X)$$

is clear. On the other hand, it is easily seen that $\mathfrak{I}_{n+t}(X)$ is generated by the set of minors of Y of order $\geq n$. All such minors belong to $\mathfrak{I}_n(Y)$.   □

Lemma 11.8. *Suppose A is a local ring and that $f : E \to F$ is a linear map as above. If $n \in \mathbb{N}$ is such that $\mathfrak{I}_n(f) = A$ then we can decompose f*

$$f' \oplus 1 : E' \oplus A^n \to F' \oplus A^n,$$

where $f' : E' \to F'$ is a linear map of free modules.

Proof. Let X be a matrix for f. Show that elementary operations on X will reduce X to the form

$$X = \begin{pmatrix} I_t & 0 \\ 0 & Y \end{pmatrix}.$$ □

In the following

$$E \xrightarrow{\ f\ } F \xrightarrow{\ g\ } G$$

denotes linear maps between finitely generated free modules with $gf = 0$.

Proposition 11.9. *For a module $M \neq 0$*

$$\underline{\mathrm{rank}}(f, M) + \mathrm{rank}(g, M) \leq \mathrm{rank}_A(F).$$

Proof. Let $0 \leq m \leq \mathrm{rank}_A(F)$ and put $n = \mathrm{rank}_A(F) - m$. We are going to prove that for some $p \in \mathbb{N}$

$$\mathfrak{I}_m(f)^p \mathfrak{I}_{n+1}(g) = 0.$$

To do this, let X be a matrix for f and d a m-minor of X. Since $\mathfrak{I}_m(f)$ is generated be finitely many minors it will suffice to show that some power of d annihilates $\mathfrak{I}_{n+1}(g)$. We can now localize with respect to d. Thus we may assume $\mathfrak{I}_m(f) = A$ and prove $\mathfrak{I}_{n+1}(g) = 0$ under that assumption. We may further assume that A is local. By 11.8 we can decompose f

$$f = f' \oplus 1 : E' \oplus A^m \to F' \oplus A^m.$$

Since $gf = 0$, we can factor g

$$F' \oplus A^m \xrightarrow{\ p\ } F' \xrightarrow{\ h\ } G.$$

We have $\operatorname{rank}_A(F') = \operatorname{rank}_A(F) - m = n$, and therefore $\mathfrak{I}_{n+1}(g) = 0$.

To conclude the proof take $m = \underline{\operatorname{rank}}(f, M)$. We have a $p \in \mathbb{N}$ such that

$$\mathfrak{I}_m(f)^p \mathfrak{I}_{n+1}(g)M = 0.$$

Since $\Gamma_{\mathfrak{I}_m(f)}(M) = 0$ we conclude $\mathfrak{I}_{n+1}(g)M = 0$ which means

$$\operatorname{rank}(g, M) \le n. \qquad \square$$

Proposition 11.10. *Suppose $M \ne 0$ is a module such that*

$$E \otimes_A M \xrightarrow{\ f \otimes 1\ } F \otimes_A M \xrightarrow{\ g \otimes 1\ } G \otimes_A M$$

is exact. Put $m = \operatorname{rank}(f, M)$ and $n = \operatorname{rank}_A(F) - m$. Then

$$\Gamma_{\mathfrak{I}_n(g)}(M) \subseteq \Gamma_{\mathfrak{I}_m(f)}(M).$$

Proof. As in the proof of 11.9 it will suffice to treat the case where A is local and $\mathfrak{I}_m(f) = A$. Since $m = \operatorname{rank}(f, M)$, we have $\mathfrak{I}_{m+1}(f)M = 0$, thus with the notation of the proof of 11.9, we get from 11.7 that $\mathfrak{I}_1(f') = \mathfrak{I}_{m+1}(f)$ annihilates M. This means that $f' \otimes 1 : E' \otimes_A M \to F' \otimes_A M$ is zero. This implies that

$$E \otimes_A M \xrightarrow{\ f \otimes 1\ } F \otimes_A M \xrightarrow{\ p \otimes 1\ } F' \otimes_A M \longrightarrow 0$$

is exact, and whence that

$$h \otimes 1 : F' \otimes_A M \to G \otimes_A M$$

is an injection. We have $\operatorname{rank}_A(F') = \operatorname{rank}_A(F) - m = n$. Notice that $\mathfrak{I}_n(g) = \mathfrak{I}_n(h)$, thus by McCoy's theorem $\Gamma_{\mathfrak{I}_n(g)}(M) = 0$. $\qquad \square$

Proof. 11.6. We keep the notation of 11.10. By assumption

$$m = \text{rank}(f, M) = \underline{\text{rank}}(f, M)$$

In particular $\Gamma_{\mathfrak{I}_m(f)}(M) = 0$. Thus by 11.10 $\Gamma_{\mathfrak{I}_n(g)}(M) = 0$, which means $n \leq \underline{\text{rank}}(g, M)$, or

$$\text{rank}_A(F) \leq \text{rank}(f, M) + \underline{\text{rank}}(g, M).$$

On the other hand, by 11.9

$$\text{rank}_A(F) \geq \underline{\text{rank}}(f, M) + \text{rank}(g, M).$$

Recall that $\text{rank}(f, M) = \underline{\text{rank}}(f, M)$ and that $\underline{\text{rank}}(g, M) \leq \text{rank}(g, M)$, we conclude therefore $\underline{\text{rank}}(g, M) = \text{rank}(g, M)$ and $\text{rank}_A(F) = \text{rank}(f, M) + \text{rank}(g, M)$. $\qquad\square$

11.3   The Eisenbud–Buchsbaum criterion

In this section we consider a complex

$$0 \longrightarrow L_s \xrightarrow{\;\partial_s\;} L_{s-1} \longrightarrow \cdots \longrightarrow L_0 \longrightarrow 0$$

of finitely generated free modules. For $i \in \mathbb{N}$ consider the i'th *partial Euler characteristic*

$$\chi_i = \text{rank}_A(L_i) - \text{rank}_A(L_{i+1}) + \text{rank}_A(L_{i+2}) - \cdots$$

Proposition 11.11. *If there exists a module $M \neq 0$, such that $L. \otimes_A M$ is **acyclic** (i.e., such that $H_j(L. \otimes_A M) = 0$ for $j \neq 0$) then all the partial Euler characteristics $\chi_1, \chi_2, \ldots, \chi_s$ are nonnegative.*

Proof. It follows from 11.6 that

$$\chi_i = \underline{\text{rank}}(\partial_i, M) = \text{rank}(\partial_i, M). \qquad\square$$

Definition 11.12. Let $L.$ be a complex as above, whose partial Euler characteristics $\chi_1, \chi_2, \ldots, \chi_s$ are nonnegative. Then define the *characteristic ideals* for $L.$ by, $j = 1, \ldots, s$,

$$\mathfrak{I}_j(L.) = \mathfrak{I}_{\chi_j}(\partial_j).$$

Remark 11.13. Suppose A is local and that $1 \leq j \leq s$ is such that

$$\mathfrak{I}_s(L.) = \mathfrak{I}_{s-1}(L.) = \cdots = \mathfrak{I}_j(L.) = A.$$

Then $H_s(L.) = H_{s-1}(L.) = \cdots = H_j(L.) = 0$ and $L'_{j-1} = \mathrm{Cok}(\partial_j)$ is a free module. Moreover, the complex

$$0 \longrightarrow L'_{j-1} \longrightarrow L_{j-2} \longrightarrow \cdots \longrightarrow L_0 \longrightarrow 0$$

has characteristic ideals

$$\mathfrak{I}_1(L.) = \mathfrak{I}_2(L.) = \cdots = \mathfrak{I}_{j-1}(L.).$$

This is easily seen by 11.7 and 11.8.

Theorem 11.14. *Let $L.$ be a complex whose partial Euler characteristics are all nonnegative, and let*

$$\mathfrak{I}_1, \ldots, \mathfrak{I}_s$$

denote its characteristic ideals. For a module $M \neq 0$, $L. \otimes_A M$ is acyclic if and only if

$$\mathrm{depth}_{\mathfrak{I}_j}(M) \geq j$$

for all $j = 1, \ldots, s$.

Proof. The case $M = 0$ is trivial, so assume $M \neq 0$ and $L. \otimes_A M$ is acyclic. Then by 11.6 for $j = 1, \ldots, s$

$$\underline{\mathrm{rank}}(\partial_j, M) = \mathrm{rank}(\partial_j, M) = \chi_j.$$

In particular $\Gamma_{\mathfrak{I}_j}(M) = 0$. Choose $a_j \in \mathfrak{I}_j$ a nonzero divisor on M. Let $a = a_1 \ldots a_s$ and let $L'.$ denote the truncated and shifted complex

$$0 \longrightarrow L_s \longrightarrow \cdots \longrightarrow L_1 \longrightarrow 0.$$

It is easily seen that $L'. \otimes_A M/aM$ is acyclic. By induction on s we have

$$\mathrm{depth}_{\mathfrak{I}_j}(M/aM) \geq j - 1$$

for $j = 2, \ldots, s$. We have already seen that $\Gamma_{\mathfrak{I}_1}(M) = 0$, thus the inequality above is also valid for $j = 1$. Conversely, suppose that the depth inequality above holds. We shall prove that $L. \otimes_A M$ is acyclic. It suffices to check the case where A is local with maximal ideal $\mathfrak{m}$. In this case the proof is by induction on s. So by the induction hypothesis we may assume

(1) $\mathrm{Supp}(H_i(L. \otimes_A M)) \subseteq \{\mathfrak{m}\}$ for $i \geq 1$.

Moreover, by 11.14 we may assume $\mathfrak{I}_s \neq A$. Since $\mathrm{depth}(M) \geq \mathrm{depth}_{\mathfrak{I}_s}(M)$, we may assume

(2) $\mathrm{depth}(M) \geq s$.

These two assumptions imply that $L. \otimes_A M$ is acyclic by 2.60. $\square$

11.4   Fitting's ideals

Let A denote a ring. Given an A-module M and a presentation by finitely generated free modules

$$E \xrightarrow{\;f\;} F \xrightarrow{\;\phi\;} M$$

define a sequence of ideals

$$\mathfrak{F}_0(M) \subseteq \cdots \subseteq \mathfrak{F}_t \subseteq \cdots$$

by, $n = \mathrm{rank}_A(F)$,

$$\mathfrak{F}_t(M) = \mathfrak{I}_{n-t}(f).$$

Proposition 11.15. *The sequence of ideals above is independent of the presentation of M.*

Proof. Let us first remark that if $\psi : S \to M$ is surjection from a finitely generated free module S, then $\mathrm{Ker}(\psi)$ is finitely generated. To see this choose a linear map $\pi : Q \to S$ such that $\psi\pi = \phi$ and consider the commutative diagram:

$$
\begin{array}{ccccccc}
P & \longrightarrow & Q & \longrightarrow & M & \longrightarrow & 0 \\
\downarrow & & \downarrow{\scriptstyle \pi} & & \| & & \\
0 \longrightarrow & \mathrm{ker}(\psi) & \longrightarrow & S & \longrightarrow & M &
\end{array}
$$

From the snake lemma we deduce an exact sequence

$$P \longrightarrow \mathrm{Ker}(\psi) \longrightarrow \mathrm{Cok}(\pi) \longrightarrow 0$$

which establishes the finite generation of $\mathrm{Ker}(\psi)$. Suppose given a second presentation of M by finitely generated free modules

$$R \xrightarrow{\;g\;} S \xrightarrow{\;\psi\;} M \longrightarrow 0.$$

Consider the exact sequence

$$0 \longrightarrow \mathrm{Ker}((\phi, -\psi)) \longrightarrow Q \oplus R \longrightarrow M \longrightarrow 0.$$

By the remark above $\mathrm{Ker}((\phi, -\psi))$ is finitely generated so we choose a finitely generated free module T mapping surjectively onto this. By construction the projections $T \to Q$, $T \to R$ are surjective. Now to compare the minors arising from two presentations, we can assume that we have a commutative diagram

$$\begin{array}{ccccccc}
P & \xrightarrow{f} & Q & \xrightarrow{\phi} & M & \longrightarrow & 0 \\
\downarrow{\scriptstyle \rho} & & \downarrow{\scriptstyle \pi} & & \| & & \\
R & \xrightarrow{g} & S & \xrightarrow{\psi} & M & \longrightarrow & 0
\end{array}$$

where π is surjective. We are going to reduce the problem to the case where π is an isomorphism. Choose a retraction r to π, and consider the diagram

$$\begin{array}{ccccccc}
P \oplus S & \xrightarrow{f \oplus 1} & Q \oplus S & \xrightarrow{(\phi, 0)} & M & \longrightarrow & 0 \\
\downarrow{\scriptstyle h} & & \downarrow{\scriptstyle k} & & \| & & \\
Q \oplus R & \xrightarrow{1 \oplus g} & Q \oplus S & \xrightarrow{(0, \psi)} & M & \longrightarrow & 0
\end{array}$$

where h and k are given by

$$h(p, s) = (f(p) - r\pi f(p) + r(s), \rho(p))$$
$$k(q, s) = (q - r\pi(q) + r(s), \pi(q)).$$

We leave it to the reader to check that the diagram is commutative and that k is an isomorphism. The minors arising from f and g are the same as those arising from $f \oplus 1$ and $1 \oplus g$. We are in this way reduced to the situation

$$\begin{array}{ccccccc}
P & \xrightarrow{f} & Q & \xrightarrow{\phi} & M & \longrightarrow & 0 \\
\| & & \| & & \| & & \\
R & \xrightarrow{g} & S & \xrightarrow{\psi} & M & \longrightarrow & 0.
\end{array}$$

Let $q = \mathrm{rank}_A(Q)$. Since f and g have the same image, we factor $f = gg'$. From 11.16 below follows

$$\mathfrak{I}_t(f) \subseteq \mathfrak{I}_t(g)\mathfrak{I}_t(g') \subseteq \mathfrak{I}_t(g).$$

Similarly, $\mathfrak{I}_t(g) \subseteq \mathfrak{I}_t(f)$. $\qquad\qquad\qquad\qquad\qquad\qquad\qquad\qquad\square$

Lemma 11.16. *Consider linear maps of finitely generated free modules*

$$E \xrightarrow{\ f\ } F \xrightarrow{\ g\ } G.$$

For $t \in \mathbb{Z}$ we have

$$\mathfrak{I}_t(gf) \subseteq \mathfrak{I}_t(g)\mathfrak{I}_t(f).$$

Proof. Left to the reader. □

Definition 11.17. Given a module M of finite presentation, the sequence of ideals

$$\mathfrak{F}_0(M) \subseteq \cdots \subseteq \mathfrak{F}_t \subseteq \cdots$$

is called the *Fitting ideals* of M.

Proposition 11.18. *Let M be a module of finite presentation. If M can be generated by d elements, then*

$$(\mathrm{Ann}(M))^d \subseteq \mathfrak{F}_0(M) \subseteq \mathrm{Ann}(M).$$

Proof. Consider a presentation of M

$$A^e \xrightarrow{\ f\ } A^d \longrightarrow M \longrightarrow 0.$$

Let $a_1, \ldots, a_d \in \mathrm{Ann}(M)$. Then the endomorphism a of A^d whose matrix is diagonal with entries $a_1, \ldots, a_d$ can be factored $a = fb$, where $b : A^d \to A^e$. It follows from 11.16, that $\det(a) \in \mathfrak{I}_d(f)$, whence $a_1 \ldots a_d \in \mathfrak{F}_0(M)$. Conversely, for any d-minor c of f we can find an endomorphism g of A^d with $\det(g) = c$ and $\mathrm{Im}(g) \subseteq \mathrm{Im}(f)$. Using the cofactor equation from linear algebra

$$g \operatorname{cof}(g) = \det(g) 1_{A^d}$$

we find $\det(g)A^d \subseteq \mathrm{Im}(f)$, that is c annihilates M. □

Proposition 11.19. *Let M and N be finitely presented modules. Then for $t \in \mathbb{Z}$*

$$\mathfrak{F}_t(M \oplus N) = \sum_{i+j=t} \mathfrak{F}_i(M)\mathfrak{F}_j(N).$$

Proof. Left to the reader. □

Remark 11.20. Consider an increasing sequence of ideals

$$\mathfrak{F}_0 \subseteq \mathfrak{F}_1 \subseteq \cdots \subseteq \mathfrak{F}_d \subseteq \cdots$$

such that $\mathfrak{F}_t = A$ for some $t \in \mathbb{N}$. Consider the module

$$M = A/\mathfrak{F}_0 \oplus A/\mathfrak{F}_1 \oplus \cdots \oplus A/\mathfrak{F}_d \oplus \cdots.$$

It follows from 11.20, that for $i \in \mathbb{N}$

$$\mathfrak{F}_i(M) = \mathfrak{F}_i.$$

We deduce from this, that if A is a principal ideal domain, then any finitely generated A-module is determined by its Fitting ideals.

11.5 The Euler characteristic

In this section we consider a module P which admits a finite resolution by finitely generated free modules

$$0 \longrightarrow L_s \xrightarrow{\partial_s} L_{s-1} \longrightarrow \cdots \longrightarrow L_0 \longrightarrow P.$$

Lemma 11.21. *The integer*

$$\sum_i (-1)^i \operatorname{rank}_A(L_i)$$

is nonnegative and independent of $L_\cdot$.

Proof. Let $L'_\cdot$ denote a second resolution of P. We can find a morphism $f_\cdot : L'_\cdot \to L_\cdot$ which is a quasi-isomorphism. The mapping cone $C_\cdot$ fits into an exact sequence

$$0 \longrightarrow L'_\cdot \longrightarrow C_\cdot \longrightarrow L_\cdot[-1] \longrightarrow 0.$$

From this we get

$$\sum_i (-1)^i \operatorname{rank}_A(L'_i) = \sum_i (-1)^i \operatorname{rank}_A(L_i).$$

The nonnegativity follows as in 11.11. $\qquad\square$

Definition 11.22. The integer

$$\chi(P) = \sum_i (-1)^i \operatorname{rank}_A(L_i)$$

is called the *Euler characteristic* of P.

Proposition 11.23. *Let $L.$ be a finite resolution of P, by finitely generated free modules. The characteristic ideals $\mathfrak{I}_s, \mathfrak{I}_{s-1}, \ldots, \mathfrak{I}_1$ of the complex $L.$ are independent of $L..$*

Proof. Put $q = \operatorname{rank}_A(L_0)$, we have

$$\chi(P) = q - \chi_1(L.).$$

From this we get for the Fitting ideal

$$\mathfrak{F}_{\chi(P)}(P) = \mathfrak{I}_1(L.).$$

This proves that $\mathfrak{i}_1(L.)$ is independent of $L..$ Consider a second resolution

$$0 \longrightarrow L'_r \xrightarrow{\;\partial'_s\;} L'_{s-1} \longrightarrow \cdots \longrightarrow L'_0 \longrightarrow P.$$

The complexes

$$0 \longrightarrow L_s \oplus L'_0 \xrightarrow{\;\partial_s \oplus 1\;} L_{s-1} \oplus L'_0 \longrightarrow \cdots \longrightarrow L_1 \oplus L'_0 \longrightarrow 0$$

$$0 \longrightarrow L'_r \oplus L_0 \xrightarrow{\;\partial'_s \oplus 1\;} L'_{r-1} \oplus L_0 \longrightarrow \cdots \longrightarrow L'_1 \oplus L_0 \longrightarrow 0$$

are resolutions of $\operatorname{Im}(\partial_1) \oplus L'_0$ and $\operatorname{Im}(\partial'_1) \oplus L_0$, which are isomorphic by 11.24 below. $\qquad\square$

Lemma 11.24 (Schanuel's lemma). *Consider exact sequences*

$$0 \longrightarrow K \longrightarrow P \longrightarrow M \longrightarrow 0$$
$$0 \longrightarrow L \longrightarrow Q \longrightarrow M \longrightarrow 0$$

where P and Q are projective modules. Then we have isomorphism

$$K \oplus Q \simeq L \oplus P.$$

Proof. Put $R = \mathrm{Ker}((P \to M, -Q \to M)) \subseteq P \oplus Q$. We establish an exact diagram:

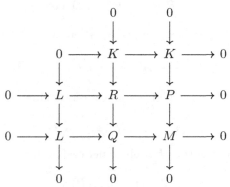

Since P and Q are projective, the middle row and column splits. Thus

$$K \oplus Q \simeq R \simeq L \oplus P. \qquad \square$$

Proposition 11.25. *Let χ be the Euler characteristic of P. For the Fitting ideals of P*

$$\mathfrak{F}_0 \subseteq \mathfrak{F}_1 \subseteq \cdots \subseteq \mathfrak{F}_d \subseteq \cdots$$

we have

$$\mathfrak{F}_0 = \mathfrak{F}_1 = \cdots = \mathfrak{F}_{\chi-1} = 0$$

and

$$\Gamma_{\mathfrak{F}_\chi}(A) = 0.$$

Proof. Consider a finite resolution $L.$ of P.

$$0 \longrightarrow L_s \xrightarrow{\ \partial_s\ } L_{s-1} \longrightarrow \cdots \longrightarrow L_0 \longrightarrow 0.$$

Let $\chi_1 = \mathrm{rank}_A(L_1) - \mathrm{rank}_A(L_2) + \mathrm{rank}_A(L_3) - \cdots$ be the partial Euler-characteristic, section 11.3. By iterated use of 11.6 we get

$$\underline{\mathrm{rank}}(\partial_1, A) = \mathrm{rank}(\partial_1, A) = \chi_1.$$

This means $\mathfrak{I}_t(\partial_1) = 0$ for $t > \chi_1$ and $\Gamma_{\mathfrak{I}_1}(A) = 0$. On the other hand, we have $\chi = \mathrm{rank}_A(L_0) - \chi_1$. $\qquad \square$

We shall summarize the properties of the characteristic ideals

$$\mathfrak{I}_s(P), \mathfrak{I}_{s-1}(P), \ldots, \mathfrak{I}_1(P).$$

To calculate $\mathfrak{I}_1(P)$ one chooses a presentation of P

$$A^q \xrightarrow{\ f\ } A^p \longrightarrow P \longrightarrow 0$$

and calculates $\mathfrak{I}_1(P)$ as the highest nonvanishing minor of f. In general one has: Let

$$0 \longrightarrow Q \longrightarrow L \longrightarrow P \longrightarrow 0$$

be an exact sequence with L finitely generated and free, then for $t \geq 1$

$$\mathfrak{I}_t(Q) = \mathfrak{I}_{t+1}(P).$$

From this description follows

$$V(\mathfrak{I}_s(P)) \subseteq V(\mathfrak{I}_{s-1}(P)) \subseteq \cdots \subseteq V(\mathfrak{I}_1(P)).$$

From now on we shall assume that A is a noetherian ring. In this case it follows from 11.21 that $\mathfrak{F}_\chi(P)$ contains a nonzero divisor.

Proposition 11.26. *Let $P \neq 0$ be a module which admits a finite free resolution.*

(1) *$\chi(P) = 0$ if and only if P is annihilated by a nonzero divisor.*

(2) *If $\chi(P) \neq 0$ then*

$$\mathrm{Supp}(P) = \mathrm{Spec}(A).$$

Proof. Let $\mathfrak{q} \in \mathrm{Ass}(A)$. If $\mathfrak{q} \in \mathrm{Supp}(P)$ then Auslander–Buchsbaum's formula

$$\mathrm{proj\,dim}(P_\mathfrak{q}) + \mathrm{depth}(P_\mathfrak{q}) = \mathrm{depth}(A_\mathfrak{q})$$

shows that $P_\mathfrak{q}$ is a free $A_\mathfrak{q}$-module. Thus, if $\chi \neq 0$, we have $\mathrm{Ass}(A) \subseteq \mathrm{Supp}(P)$. Suppose $\chi = 0$, then

$$\mathrm{Ann}(P) \subseteq \bigcup_{\mathfrak{p} \in \mathrm{Ass}(A)} \mathfrak{p}$$

is equivalent to $\mathfrak{p} \in \mathrm{Supp}(P)$ for some $\mathfrak{p} \in \mathrm{Ass}(A)$, which we have seen is not the case, and whence $\mathrm{Ann}(P)$ contains a nonzero divisor on A. $\square$

11.6 McRae's invariant

A module M is called *elementary* if there exists a finitely generated free module L and an exact sequence

$$0 \longrightarrow L \xrightarrow{\ f\ } L \longrightarrow M \longrightarrow 0.$$

The determinant $\det(f)$ is a nonzero divisor in A, as it follows from McCoy's theorem 11.2. Up to a multiplication with an element in $U(A)$, $\det(f)$ depends only on M.

Definition 11.27. Let K denote the ring of fractions with respect to all nonzero divisors. The group of *principal divisors* of A is

$$\mathrm{Pd}(A) = U(K)/U(A).$$

For an elementary module M, the invariant we have just considered will be denoted

$$\det(M) \in \mathrm{Pd}(A)$$

and will be called the *McRae invariant*.

Lemma 11.28. *For an exact sequence of elementary modules*

$$0 \longrightarrow N \longrightarrow M \longrightarrow P \longrightarrow 0$$

we have the equality in $\mathrm{Pd}(A)$

$$\det(M) = \det(N)\det(P).$$

Proof. Left to the reader. $\qquad\square$

Lemma 11.29. *Let P be a module which has a finite resolution by finitely generated free modules. If P is annihilated by a nonzero divisor in A, then P admits a finite resolution by elementary modules.*

Proof. We shall first prove that we can find an exact sequence

$$0 \longrightarrow R \longrightarrow M \longrightarrow P \longrightarrow 0$$

where M is elementary. Choose a surjection $L \to P$ where L is a finitely generated free module. If $a \in \mathrm{Ann}(P)$ is a nonzero divisor, we can take $M = L/aL$. Let us remark that R has a finite free resolution. To see this choose a finite free resolution $L.$ resp. $F.$ of M resp. P and let $f. : L. \to M.$ lift the map $M \to P$. The mapping cone $C.(f.)$ is easily seen to furnish a finite free resolution of R. Clearly R is annihilated by a nonzero divisor. We can now proceed by induction on the projective dimension of P. The construction above shows that it suffices to treat the case $\operatorname{proj dim}(P) \leq 1$. In this case, choose a presentation

$$0 \longrightarrow L_1 \longrightarrow L_0 \longrightarrow P \longrightarrow 0$$

where L_0 is finitely generated and free. It follows that L_1 is finitely generated and projective. Use a finite free resolution of L_1 to see that there exists a $p \in \mathbb{N}$ such that $A^p \oplus L_1$ is free. We have the following resolution of P

$$0 \longrightarrow A^p \oplus L_1 \longrightarrow A^p \oplus L_0 \longrightarrow P \longrightarrow 0.$$

Since P is anninilated by a nonzero divisor, it follows that the two free modules above have the same rank, that is P is elementary.   □

Let now P be a module which admits a finite resolution by finitely generated free modules and which is annihilated by a nonzero divisor. If $M. \to P$ is a finite resolution by elementary modules, we are going to prove the following proposition.

Proposition 11.30.

$$\det(M_0) \det(M_1)^{-1} \det(M_2) \cdots \in \operatorname{Pd}(A)$$

is independent of the choice of resolution $M..$

First a preparation.

Lemma 11.31. *Given a linear map $f : Q \to P$ of modules which admits finite resolution by finitely generated free modules and which is annihilated by a nonzero divisor. Given a finite resolution $M. \to P$ by elementary modules, then there exists a finite resolution $N. \to Q$ by elementary modules and a commutative diagram:*

$$
\begin{array}{ccc}
N. & \longrightarrow & Q \\
\downarrow & & \downarrow \\
M. & \longrightarrow & P
\end{array}
$$

Proof. Construct the diagram:

$$
\begin{array}{ccccccccc}
N'_{p+1} & \longrightarrow & \mathrm{Ker}(\partial'_p) & \longrightarrow & N_p & \xrightarrow{\partial_p} & \cdots \longrightarrow & N_0 & \longrightarrow Q \longrightarrow 0 \\
\downarrow & & \downarrow & & \downarrow{\scriptstyle f_p} & & & \downarrow{\scriptstyle f_0} & \downarrow \\
M_{p+1} & \longrightarrow & \mathrm{Ker}(\partial_p) & \longrightarrow & M_p & \xrightarrow{\partial_p} & \cdots \longrightarrow & M_0 & \longrightarrow Q \longrightarrow 0,
\end{array}
$$

where N'_{p+1} is determined by the exact sequence:

$$
0 \longrightarrow N'_{p+1} \longrightarrow \mathrm{Ker}(\partial'_p) \oplus M_{p+1} \longrightarrow \mathrm{Ker}(\partial_p) \longrightarrow 0.
$$

This shows that N'_{p+1} has a finite free resolution and that it is annihilated by a nonzero divisor. Choose an elementary module N_{p+1} which maps surjectively onto N'_{p+1}. In case $M_i = 0$ for $i \geq p + 1$, choose a finite elementary resolution of of N'_{p+1}. $\qquad\square$

Proof of Proposition 11.30. Let now $M.$ and $N.$ be two elementary resolutions of P. Apply 11.31 to the diagonal morphism $P \to P \oplus P$ and the resolution $M. \oplus N.$ of $P \oplus P$, to find an elementary resolution $L.$ of P and a morphism $L. \to M. \oplus N.$. This shows that any two elementary resolutions can be dominated by a third $L.$. Let us now prove that the element in 11.30 is independent of resolution $M.$ chosen. It suffices to treat the case where $N. \to P$ is a resolution and there exists a morphism of resolutions $f. : M. \to N.$. Consider the mapping cone $C.(f.)$ and the exact sequence

$$
0 \longrightarrow M. \longrightarrow C.(f.) \longrightarrow N.[1] \longrightarrow 0.
$$

The result follows from 11.28. $\qquad\square$

Definition 11.32. Let P be a module which admits finite resolution by finitely generated free modules and which is annihilated by a nonzero divisor. Choose a finite resolution $M. \to P$ of elementary modules. The following element

$$
\det(P) = \det(M_0) \det(M_1)^{-1} \det(M_2) \cdots \in \mathrm{Pd}(A)
$$

is called the *McRae invariant* of P.

11.7 The integral character of McRae's invariant

In this section we shall assume that A is a noetherian ring. By a *principal ideal* we shall understand an ideal in A, which is generated by a nonzero divisor. Let $\mathrm{Pi}(A)$ denote the multiplicative monoid of principal ideals. We have an inclusion

$$\mathrm{Pi}(A) \subseteq \mathrm{Pd}(A).$$

This identifies $\mathrm{Pd}(A)$ with the group associated to the monoid $\mathrm{Pi}(A)$. Recall, 11.26, that if P is a module with a finite free resolution, then $\chi(P) = 0$ if and only if P is annihilated by a nonzero divisor.

Proposition 11.33. *Let P be a module with a finite free resolution and such that $\chi(P) = 0$. Then the McRae invariant $\det(P) \in \mathrm{Pd}(A)$ is an integral ideal, i.e., $\det(P) \in \mathrm{Pi}(A)$.*

Proof. Let $\det(P)$ be represented $\frac{s}{t} \in U(K)$, where s, t are nonzero divisors in A. We are going to prove that $(s) \subseteq (t)$. If not, choose $\mathfrak{p} \in \mathrm{Ass}((s,t)/(t)) \subseteq \mathrm{Ass}(A/(t))$. Thus we have $\mathrm{depth}(A_{\mathfrak{p}}) = 1$. It follows that $P_{\mathfrak{p}}$ has projective dimension 1 as $A_{\mathfrak{p}}$-module, i.e., $P_{\mathfrak{p}}$ is elementary. But then $\det(P_{\mathfrak{p}}) = \frac{s}{t} \in A_{\mathfrak{p}}$ so $((s,t)/(t))_{\mathfrak{p}} = 0$ contradicting the choice of $\mathfrak{p}$. $\square$

Proposition 11.34. *Let $\mathfrak{I}$ be an ideal in A such that $A/\mathfrak{I}$ has a finite free resolution. If $\mathfrak{I} \neq 0$, then $\chi(A/\mathfrak{I}) = 0$ and $\mathfrak{I}$ contains the McRae invariant $\det(A/\mathfrak{I})$. In fact the McRae invariant is the smallest principal ideal contained in $\mathfrak{I}$.*

Proof. Let $\mathfrak{I} = (a_1, \ldots, a_p)$. It follows from 11.26, that $\chi(A/\mathfrak{I}) = 0$. Let $s \in A$ be a nonzero divisor representing the McRae invariant. $\square$

Professor Iversen's manuscript stops here. The reader is invited to finish the arguments.

Bibliography

A. Altman and S. Kleiman, *Introduction to Grothendieck duality theory*, Lect. Notes Math. 146, Springer, Berlin 1970.

M. Atiyah and I. Macdonald, *An introduction to commutative algebra*, Addison–Wesley, Reading 1969.

M. Auslander and D. Buchsbaum, *Homological dimension in local rings*, Trans. Amer. Math. Soc. 85 (1957), 390–405.

M. Auslander and D. Buchsbaum, *Codimension and multiplicity*, Ann. of Math. 68 (1958), 625–657.

M. Auslander, *Modules over unramified local rings*, Illinois J. Math. 5 (1961), 631–647.

H. Bass, *On the ubiquity of Gorenstein rings*, Math. Z. 82 (1963), 8–28.

N. Bourbaki, *Algébre commutative*, Hermann–Masson, Paris 1961–98.

D. Buchsbaum, *A generalized Koszul complex I*, Trans. Amer. Math. Soc. 111 (1964), 183–196.

D. Buchsbaum and D. Rim, *A generalized Koszul complex II, depth and multiplicity*, Trans. Amer. Math. Soc. 111 (1964), 197–224.

D. Buchsbaum and D. Eisenbud, *What makes a complex exact?*, J. Algebra 25 (1973), 259–268.

D. Buchsbaum and D. Eisenbud, *Some structure theorems for finite free resolutions*, Advances in Math., 12 (1974), 84–139.

H. Cartan and S. Eilenberg, *Homological algebra*, Princeton Univ. Press, Princeton 1956.

J. Eagon and D. Northcott, *Ideals defined by matrices and certain complexes associated with them*, Proc. Royal Soc. 269 (1962), 188–204.

D. Ferrand and M. Raynaud, *Fibres formelles d'un anneau local noethérien*, Ann. Sci. École Norm. Sup. 3 (1970), 295–311.

R. Fossum, *The divisor class group of a Krull domain*, Ergebnisse Math. Grenzg. 74, Springer, Berlin 1973.

R. Fossum, H.–B. Foxby, P. Griffith and I. Reiten, *Minimal injective resolutions with applications to dualizing modules and Gorenstein modules*, Inst. Hautes Études Sci. Pub. Math. 45 (1976), 193–215.

H.–B. Foxby, *On the μ_i in a minimal injective resolution, II*, Math. Scand. 41 (1977), 19–44.

H.–B. Foxby, *Bounded complexes of flat modules*, J. Pure Appl. Algebra. 15 (1979), 149–172.

H.–B. Foxby, *The MacRae invariant*, pp. 121–128 in: Commutative algebra (Durham 1981), London Math. Soc Lect. Note Ser. 72, Cambridge Univ. Press, Cambridge 1982.

W. Fulton, *Intercection theory*, Ergebnisse Math. Grenzg. (3) 2, Springer, Berlin 1984.

A. Grothendieck, *Sur quelques points d'algégre homologique*, Tôhoku Math. J. 9 (1957), 119–221.

A. Grothendieck, *Eléments de géométrie algébrique*, Inst. Hautes Études Sci. Pub. Math., Paris 1960–67.

A. Grothendieck, *Séminaire de géométrie algébrique*, Inst. Hautes Études Sci. Pub. Math., Paris 1960–67.

A. Grothendieck, *Local cohomology*, Lect. Notes Math. 41, Springer, Berlin 1966.

T. Gulliksen and O. Neegaard, *Un complexe résolvent pour certains idéaux determinantiels*, C.R. Acad. Sci. Paris 274 (1972), A16–A18.

R. Hartshorne, *Residues and duality*, Lect. Notes Math. 20, Springer, Berlin 1966.

R. Hartshorne, *Algebraic geometry*, Grad. Texts Math. 52, Springer, Berlin 1977.

J. Herzog und E. Kunz, *Der kanonische Modul eines Cohen–Macaulay Ring*, Lect. Notes Math. 238, Springer, Berlin 1971.

M. Hochster, *The equicharacteristic case of some homological conjectures on local rings*, Bull. Amer. Math. Soc. 80 (1974), 683–686.

M. Hochster, *Topics in the homological theory of modules over commutative rings*, CBMS Math. 24, Amer. Math. Soc., Providence 1975.

B. Iversen, *Generic local structure in commutative algebra*, Lect. Notes Math. 310, Springer, Berlin 1973.

B. Iversen, *Local Chern classes*, Ann. Sci. École Norm. Sup. 9 (1976), 155–169.

B. Iversen, *Amplitude inequalities*, Ann. Sci. École Norm. Sup. 10 (1977), 547–558.

B. Iversen, *Depth inequalities for complexes*, pp. 92–111 in: Algebraic geometry (Tromsø 1977), Lect. Notes Math. 687, Springer, Berlin 1978.

B. Iversen, *Cohomology of sheaves*, Universitext, Springer, Berlin 1986.

F. Knudsen and D. Mumford, *The projectivity of moduli space of stable curves. I. Preliminaries on "det" and "Div"*, Math. Scand. 52 (1983), 19–55.

S. Lichtenbaum, *On the vanishing of* Tor *in regular local rings*, Illinois J. Math. 10 (1966), 220–226.

H. Matsumura, *Commutative algebra*, Benjamin, New York 1970.

D. Mumford, *Introduction to algebraic geometry*, Red and 1 inch thick 196?. This item refers to Mumfords original notes printed at Harvard University in the 1960's. The notes was out of print until Springer published it as *The Red Book of Varieties and Schemes* in 1988. In the preceding years (and today) the original notes are highly collectable.

M. Nagata, *Local rings*, Tracts Pure Appl. Math. 13, Interscience Publ. 1962.

C. Peskine et L. Szpiro, *Dimension projective finie et cohomologie locale. Applications á la démonstration de conjectures de M. Auslander, H.Bass et A. Grothendieck*, Inst. Hautes Études Sci. Pub. Math. 42 (1973), 47–119.

C. Peskine et L. Szpiro, *Syzygies et multiplicités*, C.R. Acad. Sci Paris 278 (1974), 1421–1424.

M. Raynaud, *Anneaux locaux Henséliens*, Lect. Notes Math. 169, Springer, Berlin 1970.

P. Roberts, *Two applications of dualizing complexes over local rings*, Ann. Sci. École Norm. Sup. 9 (1976), 103–106.

P. Roberts, *Cohen–Macaulay complexes and an analytic proof of the new intersection conjecture*, J. Algebra 66 (1980), 220–225.

J.-P. Serre, *Faisceaux algébrique cohérents*, Ann. of Math. 61 (1955), 197–278.

J.-P. Serre, *Algébre locale – multiplicités*, Lect. Notes Math. 11, Springer, Berlin 1965.

I. Shafarevich, *Basic algebraic geometry*, Grundl. Math. Wiss. 213, Springer, Berlin 1974.

O. Zariski and P. Samuel, *Commutative algebra I–II*, van Nostrand, Princeton 1958–60.

Index